Sequestered Skydome Aetho-Haleness Organon

Sylvan Prosper

Sequestered Skydome Aetho-Haleness Organon

First Edition September 2025

Written by Sylvan Prosper

TABLE OF CONTENTS

CHAPTER 1

ATTACKS IN ANONYMIZING INCREMENTALLY UPDATED DATA RECORDS

1.1 Introduction

Data mining is all about digging for interesting and useful information which is known as knowledge from large databases or data sets. As several governments, many corporate organizations are maintaining the data repositories, there still is a room for analysis of data and thus also for mining processes. As the datasets hold precious knowledge, there exists a competency in publishing of datasets by multiple parties. The hospitals in California need to submit a report on every patient who is discharged from the hospital. The main idea behind this is to maintain a central repository of health care data nationwide for effective clinical decision making. But there is a need for a more secure and smoother data's exchange between owner and recipient. The Electronic Data Interchange (EDI) is one such initiative towards exchange of information between parties commercially. This is not only limited to single data owner and recipient but also applies to multiple data owners and data recipients. Extensible Mark-up Language (XML), Simple Object Access Protocol (SOAP) and Web Services Description Language (WSDL) are the technologies that assist in information sharing in this contemporary Information Technology (IT) era.

The rules enforced can make various devices to communicate among themselves even without intervention of owner of the device. For example, the smart fridge in your home can detect the stock details of food items inside of it based on RFID tags attached to items. Moreover, the smart fridge connected to internet can make suggestions to the home owner from the supplier and it also shows the other recipes that can be made. But for those recommendations, the supplier needs to have the details of food items inside your smart fridge. Sharing all this information with outsiders may lead to loss of individual privacy.

1.2 Motivation

In health care domain, the disease information of patients needs to be released as part of data publishing along with the details like age and location of the patients but not their private details like name and social identity number. If the personal details are inferred from this published data, then it is called a privacy concern of the patient. Many patients do not want to disclose their private information like their name, age and disease. The prevalent practices with regard to protection of privacy eliminate some of the columns of data set that may leak the privacy aspects. But this leads to distortion of data and may result in low utility.

1.3 What is Privacy Preserving Data Publishing (PPDP)?

We have to devise the strategies to get maximum knowledge from the dataset but without causing loss of privacy. This process of devising techniques towards the publishing of data without losing privacy is called Privacy Preserving Data Publishing (PPDP).In the general form of privacy preserving data publishing, the table form of raw data is as given below:

D (Explicit_attribute, Quasi_attributes, NonSensitive_attributes,Sensitive_attribute)

Where Explicit_attribute is collection of social identity numbers and names which uniquely identify record owners. Here quasi attributes are abbreviated as Quasi Identifier (QID). Quasi attributes are collection of attributes like age, the location that might possibly identify owners of the records. Sensitive attributes of the data consist of salary, disease, disability condition whereas the attributes that are non-sensitive consist of the details that may not fall in categories stated above. Mostly all these four kinds of attributes are non-overlapping.

Anonymization (Cox, 1980), (Dalenius, 1986) is the technique that hides the personal details of patients and retains the sensitive attributes for analysis of data. Explicit identifiers related to record owners must be eliminated. Though explicit identifiers are anonymized with, (Sweeney, 2002*a*) was still able to show a privacy risk of William Weld, a former governor of Massachusetts where the individual's zip code, sex and date of birth information, the attacker is still able to link the record of released medical dataset with that of voters' list of that state as shown in Fig.1.1. With the help of such attributes so called quasi-identifiers, 87 percentage of US population has been identified unique as per research report. Linking attacks happen by knowing quasi-identifiers. To prevent linking attacks, a new anonymized table T is formed:

T (QIDA, Non-Sensitive attributes, Sensitive attributes)

Where QIDA, the anonymized version of QID by applying anonymization operations on the original table D. There will be multiple records with same QIDA so that linking with other databases will become ambiguous like not being sure about which particular record among them needs to be linked.

1.3.1 Difference between PPDP and PPDM

Due to existence of issues related to privacy with regard to mining tools of data, a new research area known as Privacy Preserving Data Mining (PPDM) was raised in the year 2000. The primary objective behind PPDM is to hide sensitive information while mining with the traditional techniques. The main idea is to understand how to perturb the data and mine knowledge from the perturbed data. In PPDM, the solutions are in close association with the mining algorithm being considered. But in PPDP, it may not be the case and mining algorithm is not really known at the time of publishing of the data. In PPDP, it is required to preserve faithfulness of the data right at the record level; but PPDM does not follow any such property.

The differences between PPDM and PPDP are as follows:

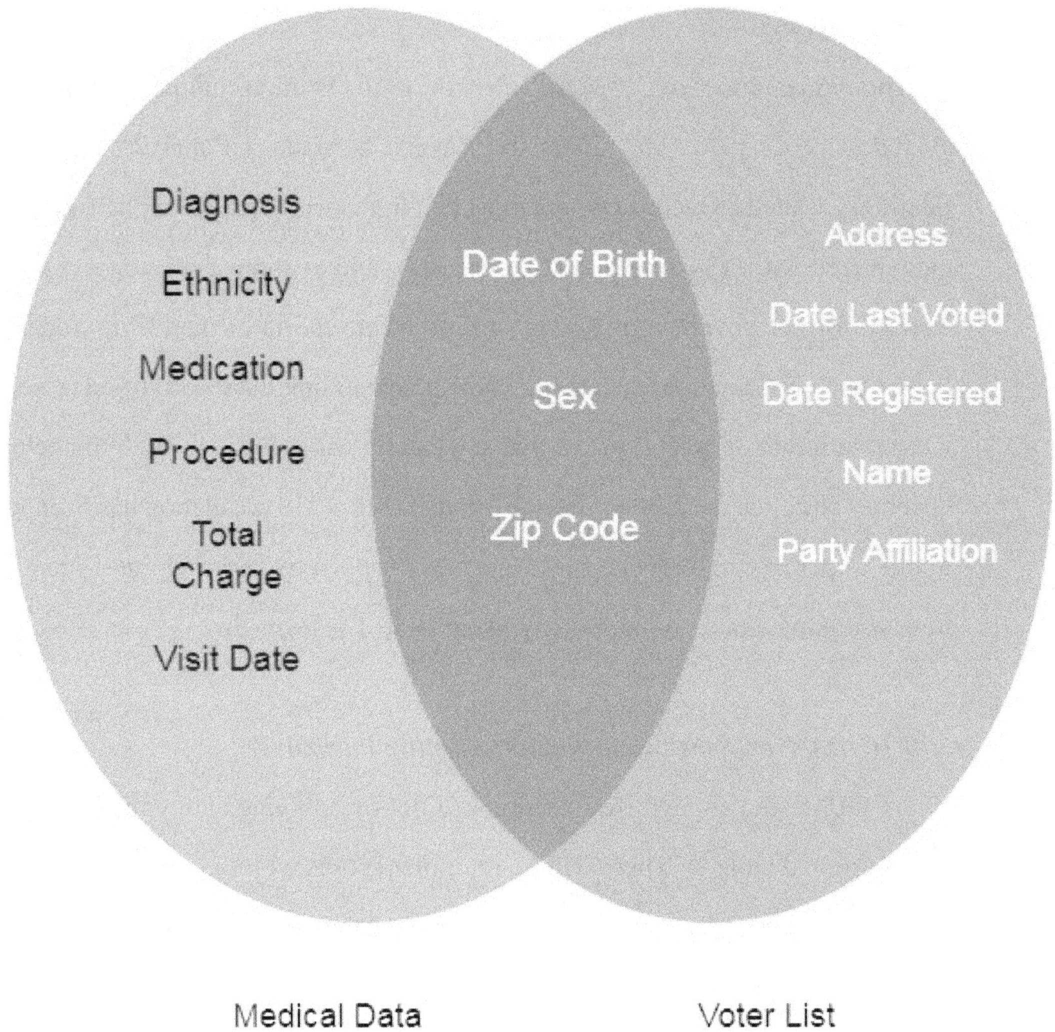

Fig. 1.1 Re-identifying owner of the record with help of voter list

a. PPDP emphasizes techniques for the purpose of publishing data but not for mining of the data.

b. Encryption as well as randomization methods may not help in preserving truthfulness of the data at the record level. So, the data release is mostly meaningless. If this is the case, the knowledge of data mining results may be released by the data holder instead of randomized data.

c. In PPDP, the details of identity of owners of the record will be hidden whereas in PPDM, the sensitive attributes themselves will be hidden.

1.4 Attack Models and Privacy Models

According to (Dalenius, 1986), any extra information should not be revealed by the published data to the adversaries regarding any victim patient even with any other background information known to them from other ways. Due to the existence of background knowledge of the adversary it is impractical to achieve absolute privacy. For example, age of the patient is sensitive information. Let us assume that the invader knows that the age of victim James is 6 years less than that of the average age of the American men. Still, the attacker did not have any idea about the average age of American men. But when the adversary can find the average age of American men from the statistical database, then James's privacy is compromised. In this context, the term victim points to the owner of record in the database. Based on the types of attacks, we categorize privacy models into two types. The first type of attack model occurs when an adversary can link the record owner to a specific record in the published table or to any other sensitive attribute in that table or probably to the entire table. These attacks are called as record linkage; attribute linkage and table linkage respectively.

The privacy models and various attack models in them are as shown in Table 1.1.

Table 1.1 Privacy Models

Privacy Model	Attack Model			
	Record linkage	Attribute linkage	Table linkage	Probabilistic attack
K-Anonymity	Yes	-	-	-
l-Diversity	Yes	Yes	-	-
t-Closeness	-	Yes	-	Yes

1.4.1 Record Linkage Model

In the model of the record linkage, the attack comes from the fact that a QID matches with a small number of records in released version T. As a result, the adversary with additional knowledge of the victim is able to infer victim's disease and will be able to identify exactly in the release T. The example below illustrates this:

The Table 1.2 is authentic patient data and Table 1.3 is the data which adversary has

Table 1.2 Original Data of Patients

Job	**Age**	**Sex**	**Disease**
Engineer	35	Male	HIV
Engineer	38	Male	HIV
Lawyer	38	Male	Hepatitis
Writer	30	Female	Fever
Writer	30	Female	Hepatitis
Dancer	30	Female	Hepatitis
Dancer	30	Female	Hepatitis

Table 1.3 Data from external source

Name	**Sex**	**Job**	**Age**
Alice	Female	Writer	30
Bob	Male	Engineer	35
Cathy	Female	Writer	30
Doug	Male	Lawyer	38
Emily	Female	Dancer	30
Fred	Male	Engineer	38
Gladys	Female	Dancer	30
Henry	Male	Lawyer	39
Irene	Female	Dancer	32

in hand from external source. Here the adversary infers the disease of Fred whose QID is $< Engineer, Male, 38 >$ from Table 1.3 and by looking at Table 1.2 only one record is matching with this QID so that Fred's disease "HIV" is leaked. Hence Fred's privacy gets compromised.

1.4.1.1 k-Anonymity

With a view to overcome the drawback of the record linkage shown in previous section, (Samarati, 2001),(Sweeney, 2002*b*), (Samarati and Sweeney, 1998) proposed the idea of k-Anonymity. According to k-Anonymity, there should be at least k records with similar QID value. The least equivalence group size on QID is k. As a result, the probability with which to link the record from the external source with that of in released version is at maximum 1/k. The data table that satisfies k-Anonymity property is called k-Anonymous.

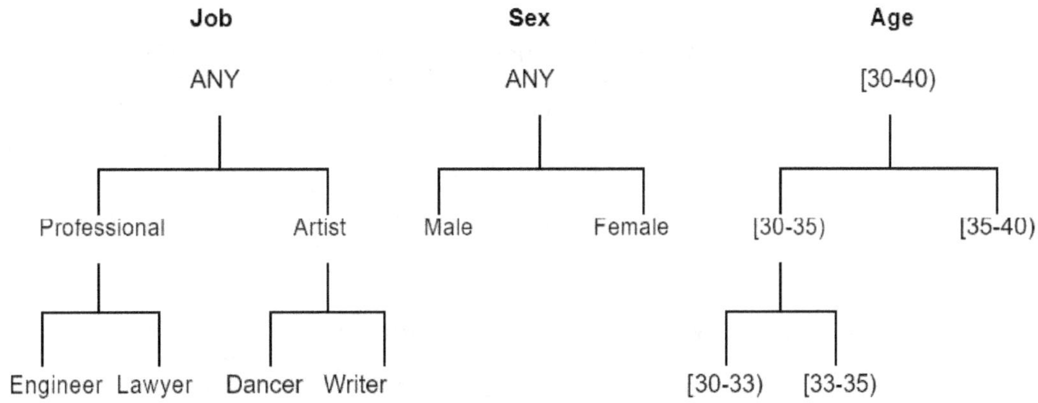

Fig. 1.2 Taxonomy trees for Job, Age, Sex

If we follow the above trees for anonymizing the columns Job, Age and Sex, the Table 1.4 comes from Table 1.2 as shown below. The Table 1.5 results from Table 1.3.

Table 1.4 anonymous table of patient data

Job	**Age**	**Sex**	**Disease**
Professional	[35-40)	Male	HIV
Professional	[35-40)	Male	HIV
Professional	[35-40)	Male	Hepatitis
Artist	[30-35)	Female	Fever
Artist	[30-35)	Female	Hepatitis
Artist	[30-35)	Female	Hepatitis
Artist	[30-35)	Female	Hepatitis

Table 1.5 anonymous data from external source

Name	**Sex**	**Job**	**Age**
Alice	Female	Artist	[30-35)
Bob	Male	Professional	[35-40)
Cathy	Female	Artist	[30-35)
Doug	Male	Professional	[35-40)
Emily	Female	Artist	[30-35)
Fred	Male	Professional	[35-40)
Gladys	Female	Artist	[30-35)
Henry	Male	Professional	[35-40)
Irene	Female	Artist	[30-35)

From the Table 1.5, we have two QIDs - namely Male, Professional, [35-40) and Female, Artist, [30-35). If the adversary knows any of these QIDs and tries to link with table-1.4's data, then there will be at least 3 records matched. So the adversary cannot disclose victim's disease confidently.

1.4.1.2 Attribute Linkage Model

Example-1.2:Let us take the Table 1.4 that is 3-anonymous.If adversary knows the QID of Gladys from external source as <Female, Artist, 30> and by seeing the Table 1.4 he can infer that 3 out of 4 tuples are having disease Hepatitis so that derived disease is Hepatitis with 0.75 probability. So Gladys's privacy got compromised regardless of correctness of disease.

1.4.1.3 l-Diversity

A new concept called l-Diversity is proposed by (Machanavajjhala *et al.*, 2007) to prevent attacks addressed in above section. l-Diversity says that every qid group should consist of at least l-well represented sensitive values. The meaning of l-well represented means that there should be l distinct sensitive values in every qid group where qid group is the collection of records having value of qid in QID.

1.4.1.4 t-Closeness

Li *et al.*, 2007 observed that when the overall distribution of sensitive attributes is skewed, i-Diversity fails to prevent attacks related to attribute linkage. For example, take the patient table where 95% of records have Fever and 5% of records have Hepatitis. Assume that a qid group has 50% of records having fever and 50% of records having Hepatitis and so it agrees on l-Diversity. But this is a major privacy threat, that with 50% confidence adversary is able to predict the disease of victim as Hepatitis even though the percentage of Hepatitis tuples is 5% in the table.

To prevent the attacks of this kind, (Li *et al.*, 2007) proposed a technique called t-Closeness. According to t-Closeness, the scattering of delicate attribute in any QID group should be close to that of sensitive attribute in whole table.

1.4.2 Table linkage model

In the record linkage and attribute linkage models, it is assumed that adversary knows the disease of victim by mapping it to external data source. A table linkage happens if the adversary can infer the presence of the victim in released table T probabilistically or confidently by linking with external data source table.

Example-1.3: Consider Table 1.4 that is 3-anonymous.Let this be T and Table 1.5 be E. As T is subset or equal to E, consider the victim Irene whose information is Female, Artist, [30-35] from E, the probability that Irene is present in T is 4 % 5=0.8. Similarly, Doug's presence in T can be inferred with 3 % 4=0.75 confidence.

1.5 Anonymization Operations

The original patient data table may not solve any privacy requirements. Hence it should be modified through anonymization operations such as generalization; suppression etc. Suppression and generalization replace the values of QID specific in nature by values of more generic nature.

1.5.1 Generalization and Suppression

Either generalization or suppression conceals some details in QID. According to taxonomy shown in Fig.1.2, the categorical attribute's specialized value may be replaced with more generalized value. In Fig.1.2, considering categorical attribute "Job", the values "Lawyer" and "Engineer" may be replaced with Professional. For a numerical attribute, the specific range of values can be replaced with more expanded range of values through intervals.

Various classes of generalization have different impacts on data utility, privacy protection and problem space. But all of them give a less specialized and consistent form of data.

Generalization replaces values of less specific in nature with values of more general in nature. The reverse of generalization is known as specialization. Suppression replaces some values of QID with a new pseudonym provided that pseudonyms are concealed. The opposite process of suppression is called disclosure. Here we study some of the generalization schemes in literature.

1.5.1.1 Full domain generalization scheme

In this scheme (LeFevre *et al.*, 2005), the attribute values are generalized to same level of the taxonomy tree. As in Fig.1.2, the values "Lawyer" and "Engineer" are anonymized to"Professional" and the values "Dancer" and "Writer" are anonymized to Artist. But search space exploration is little compared to other schemes while data distortion is much larger compared to others as the same level of granularity is needed in all paths of tree of taxonomy.

1.5.1.2 Subtree generalization scheme

In this scheme (Bayardo and Agrawal, 2005), at a node that is not leaf, either all of its children or none is generalized. For example in Fig.1.2, if the value "Lawyer" is generalized then it requires "Engineer" also to get generalized but "Dancer" and "Writer" remain ungeneralized. As a result, the generalized values form a "cut" through taxonomy tree. A cut of tree is defined as the set of values each exact one value from every root to leaf path.

1.5.1.3 Sibling generalization scheme

This technique very closely resembles the scheme of subtree generalization, but some of the siblings under a root node remain ungeneralized. As an illustration, in Fig. 1.2, if the value Lawyer is generalized to Professional, then Engineer is not generalized. This technique produces less distortion than subtree generalization as some of the child nodes which may violate threshold norms need to be generalized.

1.6 **Anonymizing Incrementally Updated Data Records**

1.6.1 Introduction

k-Anonymity technique addresses only a single static release of data, as discussed in earlier sections. But the dataset keeps on accumulating new records as new patients admit into the hospital promptly. So, latest data need to be given to researchers for effective outcomes. In the publishing of incremental databases, the data holder has previously released versions $T_1, T_2, \ldots\ldots\ldots, T_{r-1}$ and now he wants to publish version T_r where T_r is the updated version of T_{r-1} with record insertions and deletions in it. Here the difficulty is since a record for a patient remains same in all upcoming releases. Here it is assumed that all releases follow the same schema of database. There are two challenges in the problem of publishing incremental databases.

1.6.1.1 Though individual release is anonymous, the privacy of victim can be compromised by comparing different releases and eliminating some possibilities of sensitive attributes.

1.6.1.2 Another solution to solve the problem is to anonymize new records and put them as separate releases. This technique makes each release more distorted. Also, it is difficult to analyze individual anonymized releases as a whole. For example, if the Job is given as Engineer in the first month's release and as Professional in the second month's release, counting the total number of engineers is not possible. The incremental data publishing problem supposes that the invader has the knowledge of QID and timestamp of the victim. So the invader knows with certainty which release contains the record of the victim. There are two types in anonymizing data records that are incrementally updated. They are:

1) Continuous information publishing

2) Dynamic information republishing

1.7 Continuous Information Publishing

In continuous information publishing, the next release is the addition of new records to the previous release. Every new release is an accumulated version of records gathered so far in the previous releases. The k-anonymous feature of individual releases will be getting compromised, as and when the recipient cross-checks the various releases and by colluding with other recipients. These attacks are known as correspondence attacks.

Author (Fung *et al.*, 2008) put forward a technique to measure the correct number of records which can be cracked by comparison of all releases pertaining to k-anonymous function. The release may no longer be k-anonymous if cracked records are excluded. The data model for this scenario is as follows:

1.7.1 Data Model

Assume that a set of records T1 are gathered by the data holder that are timestamped as t1 and released anonymized version as R1. Then, a new set of records T2 are gathered by the data holder that are time stamped as t2 and released new version T1T2, called R2. The problem is to protect the k-anonymity of both releases R1 and R2.

1.7.2 Correspondence attacks

Consider Tables 1.6 and 1.7 with taxonomy trees in Fig.1.3 where QID is [Job, Place of Birth] and disease is a sensitive characteristic/attribute. The holder of data published 5-anonymous Table 1.6 named as R1 with records numbered a1 to a5 that are timestamped t1.The table was produced by generalizing countries like China and India just to Asia. The actual values exist inside brackets which will not be released. Next month, the data holder collects 5 new records numbered b6 to b10 and publishes 5-anonymized R2. The receiver of the data release knows that every record in R1 has a record in R2 that corresponds to it because R2 release is nothing but $T1 \bigcup T2$. Assume that the adversary tries to crack his neighbor Bob's records with the knowledge of his QID, hospitalized timestamp of admission. Consider the following three cases, each of which describes a kind of correspondence attack.

Table 1.6 Anonymized version of R1

RID	JOB	BIRTH PLACE	DISEASE
A1	Engineer	India (Hyderabad)	Hepatitis
A2	Engineer	India (Hyderabad)	Hepatitis
A3	Engineer	India (Hyderabad)	Hepatitis
A4	Engineer	India (Chennai)	Flu
A5	Engineer	India (Chennai)	Flu

Table 1.7 Anonymized version of R2

RID	JOB	BIRTH PLACE	DISEASE
B1	Professional (Engineer)	Hyderabad	Hepatitis
B2	Professional (Engineer)	Hyderabad	Hepatitis
B3	Professional (Engineer)	Hyderabad	Hepatitis
B4	Professional (Engineer)	Chennai	Flu
B5	Professional (Engineer)	Chennai	Flu
B6	Professional (Engineer)	Chennai	Flu
B7	Professional (Lawyer)	Chennai	Hepatitis
B8	Professional (Lawyer)	Chennai	Hepatitis
B9	Professional (Lawyer)	Hyderabad	Flu
B10	Professional (Engineer)	Hyderabad	Flu

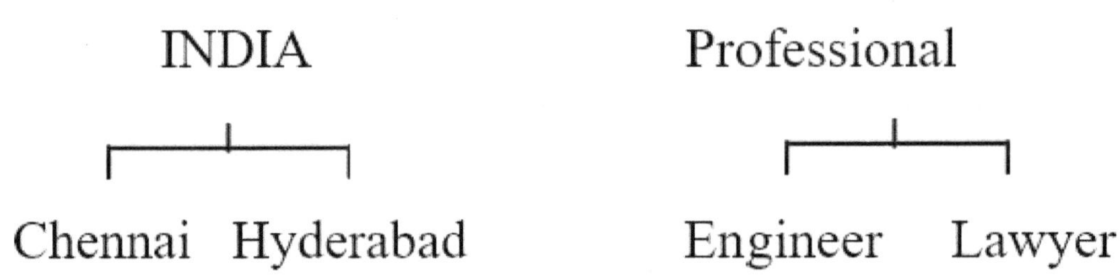

Fig. 1.3 Job's taxonomy tree and Birth Place's taxonomy tree

Case I: Assume Bob has QID= [Chennai, Engineer], timestamp t1. The adversary wants to find Bob's record in R1. Observing R1 alone; adversary can find that all 5 records match Bob's QID. But observing R1 and R2 at a time, the adversary knows that the all A1, A2, A3 records may not have been originated only from Bob's QID; or else, R2 should have contained minimum 3 records of [Chennai, Professional, Hepatitis] because each record in R1 has a matching record in R2. As a result, the adversary eliminates one of A1, A2 and A3 records. Here the selection among them is not an issue as they are the same. The crossed out record A2 is shown in Fig.1.4.

As a result, k-Anonymity of R1 gets compromised and this attack is known as F-attack or Forward Attack.

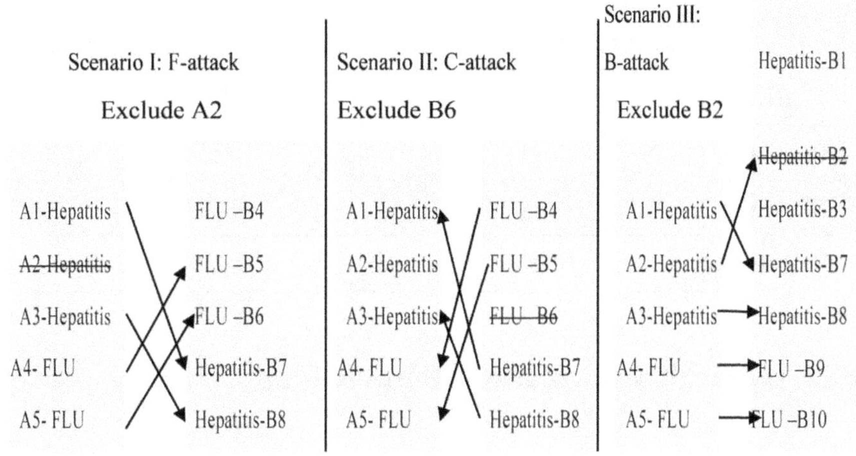

Fig. 1.4 Correspondence attacks

Case II: Assume Bob has QID=[Chennai, Engineer],timestamp t1.The adversary tries to find out Bob's record in R2 as it consists of records timestamped t1 and t2. The adversary derives that among the records B4-B8 that are matched in R2 at least one of B4, B5, B6 should contain timestamp t2, or else B4, B5, B6 would have timestamp t1 which is the case that there might have been at least 3 similar records of the form [India, Engineer, Flu].In this scenario, the adversary eliminates at least one of B4, B5, B6 since Bob has timestamp t1.The record B6 is crossed out as shown in Fig.1.4.As a result, the k-anonymity of R2 gets compromised and this attack is known as C-attack or Cross Attack.

Case III: Assume Bob has QID=[Hyderabad, Engineer], time stamp t2 and the enemy tries to find Bob's record in R2. The enemy derives that among the matching records B1, B2, B3, B9 and B10 in R2 at least one of B1, B2, B3 should have time stamp t1, or else one of A1, A2, A3 would have no similar record in R2.In this scenario at least one of B1, B2, B3 is eliminated since Bob has timestamp t2. The crossed out record B2 is shown in Fig.1.4.As a result, the k-anonymity of R2 gets compromised and this attack is known as B-attack or Backward Attack.

The types of correspondence attacks with respect to target victim's presence release, background release, and cracking release is given in the following Table 1.8.

Table 1.8 Correspondence attacks' types

	Target victim's timestamp	Background release	Cracking release
F-attack	t1	R2	R1
C-attack	t1	R1	R2
B-attack	t2	R1	R2
No attack	t2	R2	R1

1.8 Dynamic Information Republishing

1.8.1 Introduction

Xiao and Tao, 2007 has studied issues of privacy in dynamic information republishing where new records will be inserted and existing records may be deleted for every release.

1.8.2 Attacks in Dynamic Information Republishing

Assume that a hospital releases the data periodically every three months. Let Table 1.9 be release raw data T1 and tries to anonymize T1 and publishes release 2-diverse and 2-anonymous R1 as in Table 1.10. Assume that in next three months records numbered 2,3,6,8 and 10 got deleted and new ones numbered 12 to 16 got added. The resultant Table 1.11 is shown next to Table 1.10 with the new records bolded, which is noted as raw data of T2. Table 1.12 shows the 2-diverse and 2-anonymous version R2. The adversary can still infer the sensitive value of patients though R1 and R2 are 2-diverse and 2-anonymous alone.

Table 1.9 Raw data of T1

Not unleash	Quasi-identifier (QID)		Delicate (Sensitive)
Rec ID	Salary	Age	Surgery
1	22K	31	Transgender
2	24K	32	Plastic
3	28K	34	Vascular
4	35K	33	Urology
5	30K	51	Vascular
6	37K	46	Urology
7	43K	47	Transgender
8	45K	50	Vascular
9	36K	53	Urology
10	43K	62	Transgender
11	44K	66	Urology

Assume that adversary focuses on victim James and knows that his income is 22K and he is 32 years old as part of his background knowledge. From R1, the adversary can derive that James is having Transgender or Plastic Surgery. From R2, the adversary can derive that James's delicate attribute is Transgender or Urology. From the previous two derivations, the attacker can easily make out that James's surgical procedure is Transgender. This issue is known as critical absence.

1.8.3 m-Invariance

To address the problem of critical absence (Xiao and Tao, 2007) coined a privacy model called m-Invariance. The core part of this technique integrates two new concepts known as m-Invariance and counterfeited generalization. The former is a novel principle of generality and the latter is a mechanism to enforce m-Invariance if critical absence is present. As a future prospect of this work, it involves experimenting with different forms of background knowledge.

1.8.4 HD-Composition

The assumption in m-Invariance is that the sensitive attributes and QID values of a given patient do not vary over a period of time. But speaking practically, that some QID

Table 1.10 2-anonymous and 2-diverse R1

Not unleash	**Quasi-identifier (QID)**		**Delicate(Sensitive)**
Rec ID	Salary	Age	Surgery
1	[22K-24K]	[31-32]	Transgender
2	[22K-24K]	[31-32]	Plastic
3	[28K-35K]	[33-34]	Vascular
4	[28K-35K]	[33-34]	Urology
5	[30K-37K]	[46-51]	Vascular
6	[30K-37K]	[46-51]	Urology
7	[36K-45K]	[47-53]	Transgender
8	[36K-45K]	[47-53]	Vascular
9	[36K-45K]	[47-53]	Urology
11	[43K-44K]	[62-66]	Urology

Table 1.11 Raw data of T2

Not unleash	**Quasi-identifier (QID)**		**Delicate (Sensitive)**
Rec ID	Salary	Age	Surgery
1	22K	31	Transgender
4	35K	33	Urology
12	31K	35	Vascular
7	43K	47	Transgender
9	36K	53	Urology
5	30K	51	Vascular
13	40K	56	Urology
14	41K	64	Transgender
11	44K	66	Urology
15	54K	70	Urology
16	46K	75	Vascular

Table 1.12 2-diverse and 2-anonymous R2

Not unleash	Quasi-identifier (QID)		Delicate (Sensitive)
Rec ID	Salary	Age	Surgery
1	[22K-35K]	[31-33]	Transgender
4	[22K-35K]	[31-33]	Urology
12	[31K-43K]	[35-53]	Vascular
7	[31K-43K]	[35-53]	Transgender
9	[31K-43K]	[35-53]	Urology
5	[30K-40K]	[51-56]	Vascular
13	[30K-40K]	[51-56]	Urology
14	[41K-44K]	[64-66]	Transgender
11	[41K-44K]	[64-66]	Urology
15	[46K-54K]	[70-75]	Urology
16	[46K-54K]	[70-75]	Vascular

values of any patient change over a period of time. For example, if the patient moves from one place to another place the postal code changes and as a result, the analysis of patients in the old postal code area may be changed. Not only the QID value but also the delicate attribute gets changed in the course of time as a patient may recover from an old disease and acquire a new disease. To solve these issues, (Hamza and Hefny, 2013) proposed a technique called HD-Composition.

1.9 Conclusion

In current chapter, we defined the methods of data publishing and data mining pertaining to privacy preserving and described the differences between these two aspects. Besides, we defined the notion of l-Diversity, k-Anonymity, and t-Closeness as part of models of attack as well as privacy. Then, anonymization operations generalization and suppression are explored. Next, the two cases of incremental data publishing – namely continuous information publishing and dynamic data republishing are explored. As part of constant information publishing, the next release is addition of new records to earlier release and the three types of attacks in this scenario known as correspondence attacks are described. As part of dynamic data republishing, record insertions and deletions will be taken place in the existing release and as a result, the attack known as critical absence occurs. To overcome the problem of critical absence m-Invariance method is proposed. Finally, the problem of change of QID values and sensitive attributes over a period of time is attended by the technique of HD-Composition.

CHAPTER 2

CLOUD-BASED E-HEALTHCARE SYSTEM USING K-CENTROID MULTI-VIEW SIMILARITY CLUSTERING SCHEME

2.1 Introduction

Cloud computing is a combination of parallel and distributed computing models that offers incremental storage resources and various other computing resources over the internet. The basic cloud computing paradigm enables internet users to pay for their usage of resources based on pay-per-use pricing strategy. This platform also offers to scale their needed resources in both increasing and decreasing order on demand. The cloud architecture contains a diverse form of layers and sub-layers such as service layers and security or message passing in those layers. Moreover, cloud platform also offers an overall system organization that integrates all the other layers. The enhancement of cloud models supports a new set of capabilities for migrating various conservative and outdated services and applications into the cloud platform. This flexible model integrates all sorts of user domains such as developer, customer, vendors, and end-users.

In recent years, internet usage has been on the rise and has become a formidable learning and usage platform in our day-to-day lives. In this model, internet users can communicate, collaborate and share their views and suggestions through social and public forums and technologies. Similarly, the e-healthcare model offers a community service that precisely helps individuals and patients who are suffering from serious medical conditions. Furthermore, most of the time, remotely located patients are unable to get on-time face-to-face medical support by caretakers. These patients lack emotional support and are unable to reach social platform collaborations resulting from technical as well as cultural barriers.

This chapter exemplifies a privacy-preserving combined with system of e-healthcare using k-centroid clustering scheme of Multi-View similarity that seamlessly connects patients and caretakers using a common platform. This system implements a grouping mechanism that helps to communicate and address patients' health issues directly. Moreover, the patient's private information is kept confidential and will not get disclosed at any point in time. In order to achieve this privacy and integrity, a special model is adopted using k-centroid similarity algorithm of multi-view point similarity to connect and cluster all the patients' e-profiles depending on their symptom and disease similarities. The e-profiles of the patients contain personal and health status as well. All the e-profiles should undergo the hashing technique of distributed in nature to encrypt the clustered profiles in order to preserve them.

2.2 Motivation

Cloud computing is adopting many dynamic environments, even at multilingual platforms. It is quite easily accessible through the browser. However, most of the internet users are still hesitant to migrate due to security and privacy issues. Cloud platform uses virtualization model to directly meet the customer requirements. These virtual servers and data centers are accessed from remote locations. The cloud platform is not revealing information about these servers to the end-users. Moreover, privacy and security mechanisms are not entirely visible to the end-users. Though cloud offers attractive features, it may result in trustworthiness among cloud users. The computing and network resources can be availed at a cheaper cost. Data storage and data accessibility are simple through shared resources and easy to access across multiple countries without a delay. Moreover, cloud platform offers a suite of resources, including storage, application, platform, etc. The e-healthcare domain primarily requires a common shared platform to access patients' data from multiple countries. Most of the foreign countries are facing issues in implementing a remote monitoring system to check the health status of the patient. However, healthcare in developing countries is considerably worse due to lack of facilities in the hospitals located in the semi-urban and rural regions.

In general, on the internet, online communities that appear to be virtual, actually use different computing sources that demand dependable computing architectural models for rendering. But the cloud computing runs architecture that is usually a collection of service-oriented instances of integrated functionalities to support multi platform offerings. The healthcare platform applies "Software as a Service" (SaaS) for health function deployment. It happens to be proposed that privacy preservation portrays a task that definitely has a major share of the production of e-healthcare communities. The cloud is cast as a role that is integral to the inclusion of the machine for virtual execution. The style and control of healthcare communities, as well as the development of e-health profiles are not really easy as one thinks. The particulars on medical conditions are thought to be not constant and tend to keep varying. The computerized and the medical (theoretically known as Electronic Medical Record (EMR)) record is constructed by the patient through making the e-profiles of the subjects. Later, these profiles are sent for next levels of processing to similarity clustering and to connect with individual doctors. The digestive process cannot utilize the consultation of a one-on-one patient and the physician. Rather, the profiles need to be organized, and solitude measures must be employed for a particular patient. These healthcare groups connect geographically spread out doctors. The location service of inter-collaboration needs a more scalable methodology that is responsive to running patients' outlines. Making this associated with added design may well be practical with respect to the authenticity of the technology and process and place mastery for the practice of the end-users.

The cloud platform uses enlightened large applications in domain considering ease of access and the aspect that is efficient. The event between the largest utility that is accessible, like the healthcare treatment field, to the cloud is truly available. It must be noted that data privacy is a barrier which can be a serious issue in the process of further implementation. Security and privacy concerns are certainly a top priority which consists of particulars of high recordings above the business collaborations. In the recent times, the mechanisms of initial privacy preservation of the platform of cloud have been found to be missing. Privacy measures are to be taken prior to the joining and processing of data files. However, the data is usually not stable and keep varying from time to time. Privacy preservation systems must become more resilient and adequately strong to be able to work against any kind of stringent attacks. As a result of this, the cloud network is able to provide reasonable, scalable, computing facilities that are definitely easily configurable for virtual healthcare purposes, even in the case of doctors and patients not being together at a given place. The facility leads to a community–which can very well be the internet–that facilitates effort among patients, caretakers, as well as medical personnel for discussing pitfalls, relative diagnostics, and related treatment, and so on.

2.3 What is eHealth?

The term eHealth is described as "The combination of information technology and electronic communication in the health sector" by WHO. Precisely, the concept is used to ensure that the right health-related information collected or provided by the patient should be accurate and secure. Moreover, the entire information should be collected and processed through electronic media. This will help to analyze and provide needed healthcare facility to the corresponding recipient. The electronic Healthcare Record (eHR) is made use of to integrate the health-related data of a patient and describe the unique format to preserve the data in a centralized fashion. The concept of eHR is not connected or owned by any private organization or any particular healthcare provider (Van Gorp *et al.*, 2014). Moreover, the eHR is used to track the patient health status and visited healthcare organizations by the patients.

2.3.1 Significance of eHR

The eHR can be more effective, robust and can yield better solutions when these following features are incorporated:

Ease of Access – All the patients' health records and treatment-related information is in the form of digital formats. The online cloud-based systems offer enough feasibility in accessing and storing datasheets. The traditional paper-based system is very complex in terms of storage and maintenance. The in-built back-up procedures produce more efficiency when multiple records are stored at the same time.

Robust in complex environments – When integrating large scale systems along with multi-format healthcare documents, the attainment of efficiency is quite a challenging task.

High-end connectivity – The eHR creates an environment to connect and collaborate doctors with patients on a common platform. The health status documents can be shared and accessed by multiple doctors in various geographical locations with no latency.

Less expensive and fewer delivery costs – The eHR is more optimized in storing health status accurately. This will reduce multiple redundant medical lab tests for patients.

Propels to identify a new type of resolutions – The eHR is a collection of a large database that integrates patients' health records at multiple geographical locations. Every single patient has the possibility to connect with multiple doctors on a common platform. This will help patients to identify different methods of treatments and medicines as well.

Secure and Private – The entire framework is developed on a standard structural framework and well-secured access control mechanisms. This will produce more security and privacy for all patients' health records.

2.3.2 Evolution of eHRs

The national health policy is found to be complicated and more complex in attainment by the rural residents and city dwellers. The health services have been a prolonged issue which was later addressed through national level eHealth policy in the year 1990. The Australian government later implemented a standard form of eHealth policy through the National Health Information Management Advisory Council (NHIMAC) in the year 1999. Later the same policy was improved, and the scope was extended to meet specific task-oriented approaches. The core intention is to retrieve medical records among all the partners, thus reducing the communication gap between the common private and public healthcare organizations. The policy also merged with a specific scope by connecting healthcare professionals and patients during emergency conditions (HealthConnect, 2005). Later, the same policy has been reformed by the National Health and Hospitals Reform Commission. The policy ensures that all the health status documents of the patients follow the norms of personally controlled electronic health record (PCEHR) formats. This will considerably increase the efficiency and security aspects (Christy *et al.*, 2015).

2.3.3 *e*-Consent

Any medical practice should progress through legally informed consent. E-consent is a digital authentication model that allows the patients and caretakers to access their related data on a common platform. In general, the patients who are using eHealth care system give blanket consent for accessing their health-related information. This is a simple and modest methodology for healthcare organizations to handle patients and about their treatments legally. The E-consent includes multiple factors in order to maintain the records. The overall consent is a general consent form which allows granting permission for all accessibilities. However, this becomes more complicated when the patient is migrated to different healthcare organizations or physicians. The biggest disadvantage falls under this consent is to have no limit in accessing the information. Furthermore, the general consent, along with specific denials, can help patients to avoid unwanted access and also deny disclosure of their private matters (Christy *et al.*, 2015).

The general consent with limited access can contrast blanket consent that blocks all unwanted access except to the particular information. For instance, the general consent with limited access should be implemented on every new treatment initiation. Moreover, the E-consent should fulfill the following parameters before implementation.

1. Initial level patient identity verification – The patient identity information should be verified, and their health-related data should be gathered before E-consent is used.
2. Verifying that the affiliations of patients are valid – It is mandatory for the patient to have a valid membership with permitted health organization before they apply for E-consent form.
3. All the clinical conditions must be ratified, and health statement should be collected from the organization as well.
4. Verifying existing identities to avoid duplication.
5. Acquiring additional related information and understanding the purpose behind consent requests.
6. Involvement of authorized agents if needed in getting consents. The consent progress must be recorded on both allowed and denied cases.
7. Track and monitor the consent instruction connected with the health status document.
8. Consent levels should be implemented on data access. In general, the consent should follow an incremental model to define the relevant data release based on the level of consent. The data limit may have two basic categories, such as the data access on entire patient record or only for the corresponding instance of the data that can be accessed.
9. Monitor and record specific denials and typical instructions applied to complex consents.
10. Monitoring the previous data requesters and their relevant roles associated with an organization.
11. Delegation of consent migration from one organization to another.

2.4 What is Similarity Metrics?

In the chapter, the focus is placed on k-centroid similarity algorithm of multi-view point to cluster the patient's e-profiles. The privacy can be achieved through a distributed hashing algorithm. The term 'similarity metric' is used to compute the similarity factor between objects and item sets. The major application areas of similarity index are as follows:

- Basic clustering
- Correlation analysis
- NN classification, including search and prediction
- Anomaly identification
- Discrimination and classification
- Automatic categorization

2.4.1 Definition – Similarity

It is the metric to calculate the degree of two items about their identical nature. If the item sets are alike, then the degree of similarity is more. The similarity metric can be applied only on non-negative numbers. The range is generally defined as [0, 1], where zero is represented as no similarity between items and one implies total similarity between items. Some of the sample similarity measures are given below:

- Correlation-based similarity
- Jaccard coefficient
- Adjusted cosine similarity
- Extended Jaccard coefficient
- Cosine similarity
- Mean squared difference
- Dice coefficient etc.,

2.4.2 Definition – Dissimilarity

It is the metric to calculate the degree of two items about being different. The dissimilarity can be expressed with a lower value when compared to similar objects. The range value is [0, 1], where the upper range can vary from zero to infinity. Some of the sample dissimilarity measures are given below,

- Hamming distance
- Euclidean distance
- Jaccard Co-efficient similarity
- Manhattan distance
- Minkowski distance

The most common similarity metric properties can be expressed as:

1. s(a, b) = 1 (or Max Similarity) only if a = b,

2. s(a, b) = s(b, a) for all a and b, Here the similarity s(a, b) is computed via similarities between data objects a and b.

2.4.3 Cosine-based similarity

The similarity can be expressed more elaborately through Cosine based similarity. Here the consideration on any two objects or two items in the s dimensional client space. The primary calculation is applied via angle calculation to compute similarity among the items in the dimensional space. The cosine angle helps to find out the similarity between two items. The mathematical expression is formally described as n x m rating matrix that is represented as the user-item matrix. Here the similarity can be calculated for any items by considering as arbitrary i^{th} item and j^{th} item where,

$$\text{sim}(i, j) = \cos(i, j) = \frac{i.j}{||i||^2 * ||j||^2} \tag{2.1}$$

Advantage of cosine-based similarity of

• It is low-complex, particularly when sparse vectors are considered with only the non-zero dimensions.

• The real-world task is easy to implement with this measure since the time complexity of this measure is quadratic.

• The range value falls between [0, 1].

The disadvantage of Cosine based similarity

• The specification of variation rating is not included during computation. Even the rating value is not included between the users.

2.4.4 Correlation-based similarity

For the purpose of correspondence, the similitude between any two items (i.e. supposing i and j) is taken by admitting the given coalition based on likeness. In order to make coalition based computation efficient, we should limit co-evaluated parameters (i.e, scenarios where the patients have given rating to both i and j) from the very starting. If we ponder that U is the collection of patients that judged both i and j, then correspondence similitude is given by,

$$\text{sim}(i,j) = \frac{\sum u \in U(R(u,i) - *R(i))(R(u,j) - *R(j)}{\sqrt{\sum u \in U(R(u,i) - *R(i))^2}\sqrt{\sum u \in U(R(u,j) - *R(j))^2}} \tag{2.2}$$

Merits of similarity of correlation-based nature:

• The values of computation are exact.

• It may not evaluate for the profiles, and it isolates situation of where items i and j are both rated.

2.4.5 Adjusted cosine similarity

The adjusted similitude of Cosine overpowers the problem with the cosine dependent similitude. The change which may be the likeness evaluation in the client-based filter of collaborative nature and collaborative filtering of item-based nature is that in a condition the spot where the client-based CF resemblance could be computed by contemplating the rows of the matrix whereas in the event during the item-based CF the likeness is really examined through columns. The adjusted similitude of Cosine overcomes the previous weaknesses through the elimination of the users that are corresponding or item rating average from every co-rated item.

With the help of this formula, the similitude between items i and j, the adjusted similitude of Cosine is given by:

$$\operatorname{sim}(i,j) = \frac{\sum u \in U(R(u,i) - *R(u))(R(u,j) - *R(u)}{\sqrt{\sum u \in U(R(u,i) - *R(u))^2}\sqrt{\sum u \in U(R(u,j) - *R(u))^2}} \tag{2.3}$$

The benefit of adjusted cosine sameness is that:

i. It successfully overcomes all the pitfalls of similarity of Cosine.

ii. It also deducts the user's mean from each twin that is co-rated.

2.4.6 Extended Jaccard Coefficient

This extended Jaccard coefficient is always employed for the document data, and this similarity measure is condensed at the Jaccard coefficient in the eventuality of the traits, which could be digital. This co-efficient, which must be represented as EJ, is generally defined inside the equation such as:

$$EJ(x,y) = x.y/\|x\|^2 + \|y\|^2 - x.y \tag{2.4}$$

2.5 The *k*-means clustering

The *k*-means is an algorithm of centroid-located that does clustering. *'k'* presents the number of clusters, and is interpreted as an input feature. Each aspect in the particular set is consigned to a cluster center with the distance, which happens to be the smallest to this. While, the two things which might be core achieve the clustering position: 1) identify the centroids, might be or will possibly not generally be very realistic; 2) put every data element to the cluster contingent on their own distance. These two stairs iterate before the protocol converges to one best value (perhaps it certainly is nearly perfect). *k*-means looks at an algorithm which is certainly desirable and actually effortless to implement (Upmanyu *et al.*, 2010). Provided that the distance is a decent and reasonable option of k commercial value, it usually achieves logical results. Figure 2.1 depicts the procedure of the k-means that is generally standard clustering where k = 3.

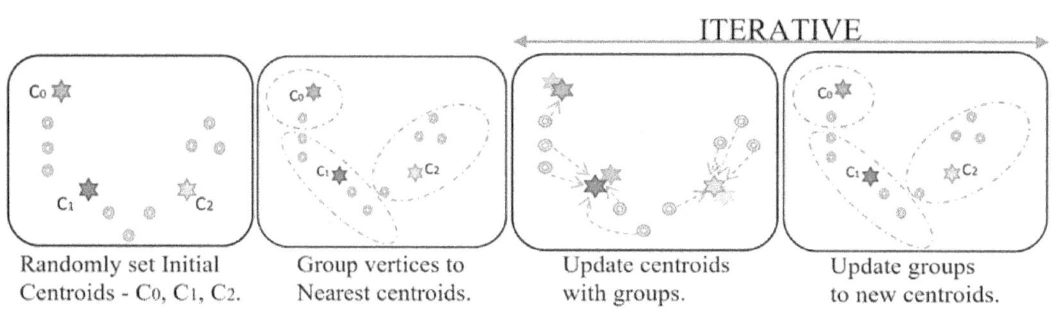

Fig. 2.1 k-means clustering process

The k-means algorithm 2.1 is given below.

2.5.1 Algorithm 2.1:The steps of the k-means algorithm

Assume that a set of profiles $\{x_1, x_2, ...x_n\}$ where each profile x_i is a vector of dimension *d (d is an integer)*; the objective is to divide the set into *k* various clusters. The different steps are involved in the k-means clustering algorithm:

1. Initialize *k* centroids

$$\{m_1,\ m_2,\ m_3, \ldots\ldots.., m_k\}$$

To make it simple, let's select them on random basis.

2. Repeat the following steps until all centroids do not change:

a).Associate an element x_i to cluster S_j, with regard to,

$$S_j = \{x_i : \|x_i - m_j\|^2 \leq \|x_i - m_t\|^2, 1 \leq t \leq k\} \tag{2.5}$$

b).Modify the centroid of S_j namely, m_j

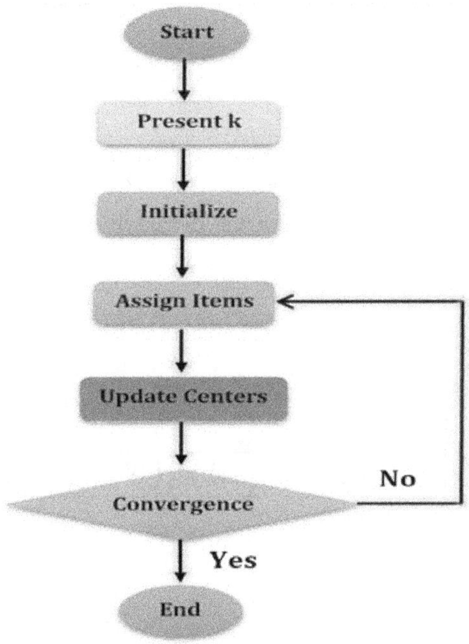

Fig. 2.2 k-means clustering flow diagram

2.5.2 *k*-means algorithm implementation

One small sized data set is used to indicate the processing steps of the k-means algorithm. Example 1: Assume 8 data points: (2, 2.4), (2.1, 2.2), (2.4, 2.7), (5.9, 6.03), (6.3, 5.8), (6.20, 6.35), (6.85, 6.25), (3.12, 7.12). We use k-means algorithm to make the set into two different clusters. The initial data set is in Table 2.1.

The state will be processed in the upcoming subsections step by step.

Table 2.1 Data set to be clustered

Object No.	State A	State B
0	2	2.4
1	2.1	2.2
2	2.4	2.7
3	5.9	6.03
4	6.3	5.8
5	6.2	6.35
6	6.85	6.25
7	3.12	7.12

2.5.2.1 *Iteration - 1*

Assume initial cluster centers:

m_1 = (2.4, 2.7),

m_2 = (5.9, 6.03).

Table 2.2 Initial k-means cluster centers

Cluster	Object No.	Centers
Custe1	2	m1: (2.4,2.7)
Cluste2	3	m2: (5.9,6.03)

Euclidean distance between centroids and all the objects will be computed, and the results are given in Table 2.3.

Each point is associated to the closest cluster. Then, the collection of points is bifurcated

Table 2.3 Euclidean distance from each object to centroids

Object No.	Distance to m1	Distance to m2
0	**0.5**	5.328
1	**0.583**	5.395
2	**0**	4.831
3	4.831	**0**
4	4.982	**0.461**
5	5.269	**0.439**
6	5.693	**0.975**
7	4.478	**2.986**

into 2 clusters:

$cluster_1 = \{0, 1, 2\}$ and $cluster_2 = \{3, 4, 5, 6, 7\}$.

Update both centroids m_1 and m_2:

Compute the updated m_1 of $cluster_1$ as given below.

$m_1 = ((2 + 2.1 + 2.4), (2.4 + 2.2 + 2.7))/3 = (2.16666, 2.43333)$

Compute the updated m_2 of $cluster_2$ as given below.

$m_2 = ((5.9 + 6.3 + 6.2 + 6.85 + 3.12), (6.03 + 5.8 + 6.35 + 6.25 + 7.12))/5 = (5.674, 6.31)$

2.5.2.2 *Iteration-2*

Updated centroids are used to decide the distances once again. Table 2.4 displays the results.

Table 2.4 Distances of Euclidean from each point to updated centroids of cluster

Obj No.	Distance between object and m1	Distance between object and m2
0	**0.16924**	5.36529
1	**0.24217**	5.44661
2	**0.35502**	4.87352
3	5.1847	**0.8074**
4	5.33165	**0.80744**
5	5.62281	**0.52751**
6	6.04229	**1.17752**
7	4.78311	**2.67936**

Each data object is associated to the closest group. Each set then is divided into the new clusters: cluster1 = 0, 1, 2, and cluster2 = 3, 4, 5, 6, 7. From both iterations, the clusters formed are same as that of earlier iteration. Therefore, we stop the process as it converges.

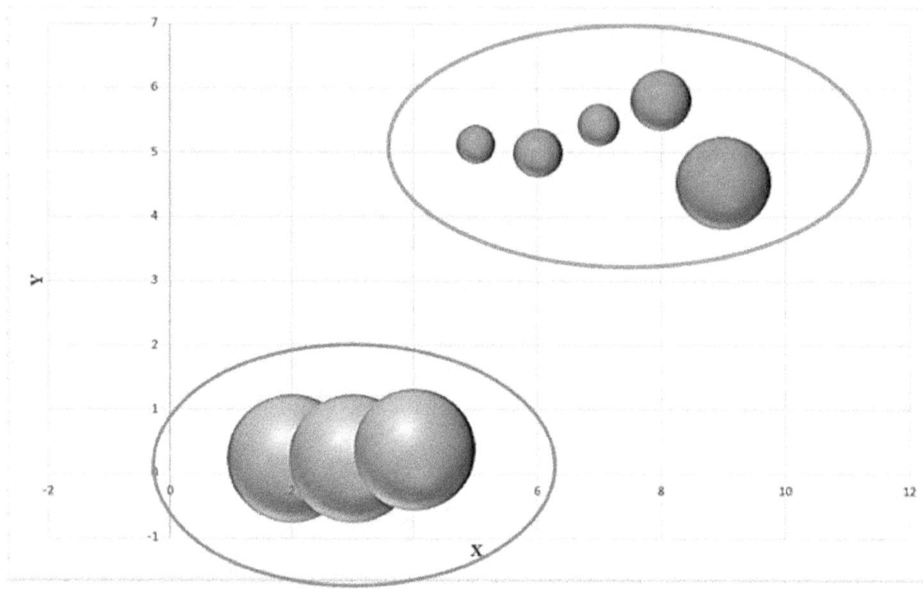

Fig. 2.3 Envisages the result

Figure 2.3 projects results of current clustering technique for data set in Table 2.1. It can be seen in Fig.2.3 that the 3 points appearing in blue are fairly close to each other; same is the case also for the 5 points in red. The results of the clustering are as expected, indicating that this method works fairly well for the given set.

2.5.3 Limitations of k-means clustering

k-means clustering performs fairly well for the information data set in the above section. Do note that this doesn't do so surely works with all forms of data sets. Quite often, this method has got certain limitations as well.

1. It also highly hinges on the group that is exactly initially and extremely when models are unable to contain many objects.

2. k, the true host of several clusters, has something to wind up being put in advance. In reality, determining 'k' that is certainly appropriate value is one of the several battles coming from clustering algorithm of k-means.

3. Given the results are usually same that it may drive different computational results after the objects are situated in new orders. This one is really a problem that could possibly get magnified as the genuine number of data snippets increase.

4. k-means grouping algorithm is unable to perceive non-spherical clusters.

2.6 Literature Review

Cloud computing techniques are fairly more efficient for any domain of intricate application. These techniques use various online applications for the tasks of data production, processing, and storage. Sharing wellness information with cloud computing is an extra important element and requires unique significance for data privacy concerns. Companies of Cloud Computing Services may obtain straight accessibility fall saved information and can market these information datasets without suitable approval. Majority of on-line applications make use of the pay-as-you-go model, and total accessibility is given on need basis. The majority of applications in healthcare domain utilize software such as Microsoft HealthVault for Cloud Computing Services. They heavily depend on information privacy, safety and security actions to deliver patient wellness details via the individual health and wellness record and format of ElectronicHealthRecord (EHR) in (Sreedhar *et al.*, 2017).

This is the most common format not only for the purpose of collecting information but also for eliminating geographical obstacles between health and wellness organizations and patients (Giniat, 2011). Sharing health data through cloud computing is a more crucial element and needs unique relevance for information personal privacy concerns. According to a previous research study (Wu, 2012), the usage of straight client-side cloud computer is about 70%, while consumption by the third-party dealers constitutes roughly 65%. Optimum intake might cause data altering and unauthorized access. As the cloud sustains open accessibility through a browser, applications that are used online are more susceptible to privacy and safety attacks. Therefore, majority of the premier health care companies are still not sure about moving to the cloud platform.

An author (Wang *et al.*, 2013*b*), (El-Gazzar *et al.*, 2016) defined that the primary obstacle to development in modern cloud computing services and technologies is the safety violation at the side of the cloud service provider. Cloud computing/storage service carriers may well be given straight accessibility to all saved information and may offer these information sets by not taking ideal permission. Likewise, this may be forecasted as expert assaults.

The cloud platform has quick handling capacities that can do any sort of complex organization tasks. Healthcare is one of the most encouraging as well as testing domains for implementing cloud services. Author (Lin *et al.*, 2014) discussed the medical care access in countries like China and offered standards for migration towards cloud technology. The expected framework uses immediate feedback and scalability and the opportunity of saving different multimedia materials. In another work, (Khan, 2012) labelled numerous artifacts versus cloud usage for medical care organizations. The author also recommended numerous affordable techniques. The recommended standards stress the expediency of using various options of cloud services for the health care domain name. In 2013, a personal company came up with a total collection of cloud gain access to a plan known as centricity 360 (Monegai, 2013). This on-line portal enables numerous clinical collaborations, assists in the organization in between caretakers and individuals, and attaches to ideal physicians for total treatment. These methods reduce repeat testing and eliminate uncalled-for low-order diagnostics for additional handling. Similarly, author (Thilakanathan *et al.*, 2014) executed some efficient analysis through utilizing less cost occurring criteria to accomplish maximum utilization and also minimal response time for cloud-based operations of long tenure. In a similar way, in one more work, (Sultan, 2014) carried out a study of highest utilization by making use of public as well as exclusive medical facilities' data transactions on a cloud platform. Both the research studies mentioned above have very well described cloud-based medical care applications. These researches have not provided any type of trustworthy structure for healthcare application use. Several IT companies have begun to assist mi-

Table 2.5 Privacy Preservation – A Comparative analysis

Investigation Manuscript	Proposed model	Pros	Cons
(Sreedhar et al., 2017)	Hybrid Technique to Preserve Privacy over Incremental data sets	Anonymises data with top- down and bottom-up approaches	Did not discuss how to collect item sets from patients in distributed data sets without losing privacy
(Wu, 2012)	Secure sharing of electronic health records in cloud	Discusses percentage of data access domain wise	Lacks in protecting privacy
(Wang et al., 2013b)	Technique to protect privacy in mobile sensing‘	Discussed techniques in mobile sensing	Not discussed privacy with regard to healthcare
(Lin et al., 2014)	Proposed standards for migration into cloud technology	Worked on protecting multimedia data	Not discussed privacy preservation over data
(Khan, 2012)	Suggested various techniques and discussed issues in migrating to cloud	Analysed the various issues broadly	Not discussed privacy preservation of data in health care
(Thilakanathan et al., 2014)	Suggested technique to achieve maximum utilization and minimum response time	Focused on utility and energy of the system	Not elaborated and discussed on privacy preservation of health care data

gration of information of medical care organizations towards cloud technology. Such organizations are yet reluctant to transfer their information to the domain of cloud.

Several authors paid attention to the security and safety constraints on the cloud system to accomplish an optimum gain of changes as given in (Singh and Liu, 2003). The cloud is a two layer design where both the customer and provider of the cloud services are the significant elements. When the data is submitted, the cloud service provider takes complete charge of information storage as well as information privacy conservation (Van Gorp *et al.*, 2014). The cloud service supplier needs to likewise guarantee information schedule and also information integrity. If the customer takes in minimal resources for jobs such as information storage, this might show information personal privacy worries "or" The minimal accessibility of the stored data increases data privacy concerns. In the information storage process, data redundancy is the most essential function. Redundancy of data suggests the undependability of the cloud solution company. Additionally, the cloud company requires to maintain the data to gain a greater credibility. The cloud framework is affordable for storage space of large amounts of data. Data personal privacy functionalities for storing data in cloud must be extra transparent at the level of both the ends. For instance, no sound assurance as well as insurance coverage has been specified for the integrity as well as accessibility of the information kept on the cloud (Rocha *et al.*, 2011). These concerns may produce higher impact on the realization of the environments of cloud. Efficient information stability and availability need optimal checking of quality issues and audit patterns to enhance the credibility (Michalas and Komninos, 2014). This may trigger financial obstacles for cloud solution companies, all the problems mentioned above related to cloud storage need to be resolved to prevent information personal privacy conservation issues (Pasupuleti *et al.*, 2016), (Wang *et al.*, 2013*a*). Attending to these concerns may play a considerable function in constructing the trustworthiness of cloud solution suppliers.

The internet social community, generally described as the online area, is defined as a neighborhood in which several individuals connect with others via a net system. Membership is called for prior to initiating any type of communication as well as providing any details to the website. These on-line neighborhoods can be predicted as understanding details of systems due to the fact that lots of members of a team will be able to interact with one another by publishing valid details and talk about different topics (Li *et al.*, 2017). Typically, anybody in the group can provide recommendations as well as team up with other discussions. Individuals can connect with e-mail applications, expertise online forums, talking website, blog sites, etc. This kind of solution helps to keep track of upcoming celebrations as well as events or any kind of sporting events. In this sector, medical care communities have extremely arisen to share details related to health and also provide appropriate recommendations as well as assistance to individuals and their center members. The internet areas may be utilized as calendars to stay on top of events or to acquire any other information related to treatment.

2.6.1 Comparative Analysis

Clustering is a partition of information sets or data objects into groups of comparable items. Signifying the data by smaller clusters always sheds particular exceptional facts, yet accomplishes generalization. It prototypes information by its clusters. Data designing positions cluster in a historical perception deep-rooted in stats, data, and statistical analysis. The concept of clustering is the substance of active study in several areas such as, Machine Learning, pattern recognition, and statistics.

At the outset, there is a massive calculation intricacy associated with typical Singular Value Decomposition (SVD). The actual calculation intricacy of SVD is quadratic in the list of document matrix through term based approach (the rank is constrained by the lesser sized of the variety of categorized records as well as the variety of terms) and cubic formation of particular values that are calculated (Sreedhar *et al.*, 2017), (Giniat, 2011). On the other angle, most of the times of SVD for a term-document matrix, the variety of records is fairly less than the variety of index terms. On the other hand, the variety of singular values that is equivalent to the term-document matrix's rank, likewise based on the number of categorized documents. Therefore, we can be said that the calculation complexity of SVD is entirely identified by the quantity of document vectors in the matrix of term-document. If categorized documents' number in the matrix of term-document is reduced, then the big calculation complication of SVD can be lowered.

Generally, clusters prevail through file collection. Typically, there are diverse subjects spread in various files via message collection. Also, if all categorized records in a cluster are related to the same topic, they can be split right into numerous subtopics. SVD has the capability to discover the most illustrative vectors for text depiction, it may not be optimum in perceptive records with diverse semiotics. In information retrieval, the pertinent records with the inquiry should be recovered as many as conceivable; on the other hand, the pointless files with the inquiry should be recovered as soon as possible. If major clusters, in which records have strictly related semantics, can be drawn out immediately, then the appropriate files can be fetched in the cluster with the assumption that carefully linked categorized documents have a tendency to be appropriate to the very same request; that is, appropriate documents are extra like one another than they are like no pertinent categorized documents.

Appropriate contextual evidence exists along with the co-incidence of given index terms in records. The method classic weighting plans (Wu, 2012), (Wang *et al.*, 2013*b*) are anticipated on the source of details about the occurrence circulation of index terms inside the whole group or within the pertinent and also no significant clusters of documents. The primary model of the term weighting patterns is a probabilistic one as well as it thinks that the index terms made use of for illustration are dispersed independently in records. Thinking variables to be autonomous is generally an issue of precise ease. However, the basic nature of details access, exploitation of dependency, or association among index terms or documents will be a certainty usually because a better access results such as the majority of direct algebra approaches suggested for LSI (Shakeabubakor *et al.*, 2015), (El-Gazzar *et al.*, 2016). That is, from technical perspective, the index terms in records are reliant on one another. In the point of view of grammars, relevant words are disposed to have burstiness in records, as well as lexical words regarding the very same topic, are probable to rebuild in the exactly same material. Therefore, the contextual terms of an index term could likewise be highlighted and also put together when utilized for retrieval. Catching the co-occurrence of index terms in documents, as well as additional capturing of the co-occurrence of documents with some typical index terms, are of immense significance to characterize the connections of records in a message collection.

In another dimension, the k-Nearest Neighbor (kNN) classifier is among one of the most typically utilized systems for supervised learning methods (Lin *et al.*, 2014). This technique simply calls for a disparity that may be pointed out in between two given instances. Essentially, kNN classifies an offered input element through appointing one of the most mutual label tags between its k-nearest models of the defined training sets specified depending on to that inconsistency. Further, most of its reputation originates from its theoretical easiness as well as simple treatment that handle amazingly in a number of pattern recognition tasks. This is actually well suited to disparity coming across multi-class classifications, this will cause when the set of feasible labels comprises more than two elements (Monegai, 2013). Keeping this sense, contrasting various other protocols, including Support Vector Machines that need to pick some type of method to readjust to this instance (Monegai, 2013). The kNN policy carries out certainly any kind of strategy customization due to its multi-class nature.

As an illustrative example of instance-based algorithm, the kNN classifier does not perform a specific generalization procedure (i.e., building a model) due to the preliminary training data but directly deliberates those examples for classification (Thilakanathan *et al.*, 2014). The classifier, as a result, progresses its efficiency as the training collection rises (i.e. Training sets), having actually been shown its uniformity as the variety of training instances of 20 methods to infinity (Sultan, 2014). Apparently, our community is strongly recognized by the huge quantity of information bordering us. Since the beginning of the information associated to modern technologies, information generation has been stated as continuously expanding (Li *et al.*, 2017), this consequence being more notable in the last few years. A kNN classifier probably can use these large information resources to boost category efficiency. Nonetheless, given that the kNN classifier tends to calculate a range in among the input sample and also every example of the training information, it necessitates reduced effectiveness in both category time and also memory use. This comprises the main downside for this above classifier, which comes to be an overwhelming challenge when taking such large significant scale of training quantities.

The primary presumption is that the k-nearest neighbors of an offered instance hinge on the equivalent cluster. Hence, the kNN pursuit can be effectively done in 2 actions: i) getting to the nearby cluster; and also ii) discovering the k-nearest neighbors inside the cluster. Considering a large-scale situation, this would ultimately conserve a significant amount of distance calculations, therefore carrying out the procedure much more successfully (Singh and Liu, 2003). Yet, there is an opportunity that this search or exploration involves some precision loss if component of the kNN falls in diverse clusters. To relieve this circumstance, a method is devised to make sure that this opportunity is a lot more not likely. Our concept is that clusters are not permanently disjoint; however, there are instances which can fit into greater side. For attaining that, we use an added action after the preliminary clustering procedure: (i) concentrating on one collection, we iterate via every one of the instances; (ii) For every component of the collection, we inspect the k-nearest neighbors allowing for the whole training set; (iii) This situation for any one of the k neighbors of the instance at issue is not component of the collection being considered, we embrace it inside the cluster, therefore potential of the area segmenting to something comparable to an unclear clustering; (iv) this procedure is provided for each of the clusters acquired. This method raises the possibility of creating all the k-nearest neighbors of an offered test circumstances autumn in the same collection. Additionally, note that the clustering procedure and the anticipated enhancement are done together as a pre-processing stage, consequently not influencing the effectiveness of the category process.

As it is intended to be checked experimentally, this procedure of boosting the cluster size connects the brute-force kNN situation with regard to accuracy or precision with far much a lesser amount of computational expense. Current advancements in feature learning, namely deep learning, actually made a development in the capability to find out appropriate features for classification. Unlikely, rather than renovating to hand-made features extracted. The described models accomplish the training to presume, out of the raw input data sets, that the most ideal features for the job helpful. This characteristic understanding is accomplished through Deep Neural Networks (DNN), being composed of a variety of layers which have the ability to represent diverse levels of abstraction from the input data sets. Some authors (Van Gorp *et al.*, 2014), (Rocha *et al.*, 2011), nonetheless, have revealed that the DNN can be just used as a feature extraction engine. That is, serving the network along with the input data sets and also taking among the intermediate or transitional depictions, the majority of the second-to-last layer output, as features for the classification tasks. It calls for the attributes and the range considered to be ample for the task. Taking into interpretation that DNN and kNN are completely balancing in terms of function removal as well as decision limits, it is motivating to recommend a hybrid system wherein both algorithms can manipulate their possible as well as see their downsides alleviated by various others. The positive side of a hybrid technique has been validated via other classifiers such as Support Vector Machines (Michalas and Komninos, 2014), yet a thorough study in the instance of large information with kNN remains open forum. It needs to be kept in mind that essential input information with one of the most suitable functions could have a substantial influence on the enactment of the clustering algorithm. Consequently, it is to be anticipated that these appropriate attributes will additionally help to enhance the estimated search of the kNN.

Keeping the above two techniques in mind, the k-means is the comfortable unsupervised intelligent algorithm which can lighten the widely known clustering issues (Wang *et al.*, 2013*a*). The technique marks after a simple and uncomplicated method to a group provided relevant data sets established by means of groups (k groups) which have been developed in advance. The basic impression is actually to illustrate k centroids, a single entity for every group. These centroids consume by a guile fashion leading in a distinctive region of dissimilar impacts. Thus, the better way is actually to position them as far as possible connection from each other. The next step is to take each factor within an offered information sets and also co-partner it along with the neighboring centroid (i.e. closest centroid) until getting to a state where the entire factors are actually related to groups. When the first phase is carried out as well as an unanticipated aggregating is actually accomplished immediately, it is needed to reconfigure a new set of k centroids as barycenter's of respective group as a result of the last measure. After making thek brand-new centroids, an additional binding has to be set up among the equivalent centroids put as well as the closest newly identified centroid. A pattern will certainly be made. As a result of this particular pattern, It could be recognized that the k centroids may certainly alter their structured locations, and also, towards the end, the centroids will certainly not transform their settings any longer (Pasupuleti *et al.*, 2016), (Nguyen *et al.*, 2011). Carefully considering, the k-means algorithm is structured as four sequential steps, and also it is complying with order. (a) Randomly keep k components or elements in a space and cluster the item coordinates. (b) Allot each item in the given space to a new group that is actually the absolute and most related to it. (c) Afterwards, the consignment of all the item coordinates in the defined space, re-compute the k centroid elements and modify their places, specifically. (d) Repeat procedures (b) as well as (c) until the centroids get to a position where they no longer transform with their distance between all the elements of their group.

Eventually, the strategy is going to definitely finish; nevertheless, the k-means estimation does not often discover the best optimum style, or even at the very least contrasting it to the least universal unbiased feature that has actually been mentioned. The algorithm or procedure is actually fragile and also basically reliant on the beginning randomly designated group centers. The algorithm procedure should perform at various times to lower this influence though it is going to select a diverse set of centroids each time, assembling it really hard to contrast its primary signs (de Assunção *et al.*, 2010) the above mentioned steps illustrate the k-means algorithm as well as its output. An appealing invention was attained by (Otair, 2013), where they recommended an alternative to k-means using a mechanism referred to as KD tree structure. This is constituted making use of a routine tree algorithm, yet it has an exclusive information framework (i.e. data structure) to hold the space or distances in between the nodes, that reduces each cyclic progress of k-means expressively. It is a relatively scalable enactment of k-means; however, the shortfall is actually sequential in attributes. The author built an algorithm depending upon k-means. Consequently, it takes as considerable data as possible also drives it into central memory. This could be a significant trouble on enormous datasets as the algorithmic procedure may certainly not accommodate every little thing into the memory. Despite the fact that the authors planned a relatively intricate compression algorithm to yield right into account, the significant information issue, it has actually not operated in a lot of the speculative examination test cases. Development from this algorithm has been presented to be actually slower than the initial k-means however will certainly resolve deadlocks created due to the initial k-means. A version to the proposed algorithm was created through to possess more density via compression and also to decrease the central memory usage. In order to possess one-dimensional array for the correlation coefficient, their idea is certainly not to support every little thing in the main memory. This can easily quicken handling, yet it is going to merely create the centroid points marks, certainly not all the given clusters.

On other hand, the hierarchical clustering formation described by the authors Zhang and Forman that constructs a tree-like framework by incrementing its own index and reprocess and rescan the data on every repetition (Forman and Zhang, 2000). Through this concept, the method BIRCH is generated, along with a data framework through Cluster Feature Tree. This particular tree is that every node has three-way features such as node variety, weight summation-squared, and also weight summation. Nevertheless, the solution delivered through the algorithmic procedure is actually an estimated remedy only. The drawback is actually that the size and scope of the tree could be tough to take care of on a big scale because if it performs unsuitable to fit into memory, after that, BIRCH becomes unfeasible. One more heuristically determined protocol is CURE (Guha *et al.*, 2001). Due to the fact that it is actually advised to be a treatment for clustering on very big data banks. It uses arbitrary random testing, and it segments the data right into smaller sized groups. After that, it begins the computation for each smaller group and also relates the groups all together. Consequently, it will certainly combine each pair of groups that consume a resemblance feature lower than a particular threshold and also will definitely grab the centroid of the brand-new groups by re-computing the mean for all the factors, precisely as k-means performs. The protocol goes on knotting up until it treasures all the equivalent sets that may not be combined any longer. The end results generated supply simply a relative response. The author suggested a heuristic procedure for out-of-core clustering (Guha *et al.*, 2001). The proposed concept is actually to divide the data in to more number of chunks that the primary memory can try and manage to scan the whole memory merely once after then executes k-means algorithmic procedure on the data chunks precisely.

The author produced a statistical methodology to parallel (i.e. similarity cluster) clustering (Guha *et al.*, 2001); the protocol is actually supported through a data design based upon dendrograms. In the proposed model, the authors try to amalgamate one coming from a neighborhood dendrogram to a universal one by each version of the algorithm to gather information as well as based on recently determined limits, the algorithm will definitely go on repeating till the limit is connected with. Since the framework of the dendrogram is continuously increasing and also at a particular factor in opportunity went out of main memory, the algorithm was tried out on huge datasets and also was certainly not effective. The importance of their approach is consuming parallel clustering method, that certainly not a single dendrogram but several other were being created at a single run; once a dendrogram connects with a specific threshold that have been specified previously, it is going to be merged along with additional ones up until they create ultimate dendrograms each of which is looked at a cluster, the centroid is made through the root of the tree-like structure. A reasonably comparable method was sought through the proposed work, beginning with the base up, but coping with distance statistics build up by utilizing an association or connotation regulation to the local data. The later one started the suggestion of assembling coming from a local (native) cluster into a global (superficial) one, taking right into factor multiple sets accumulation. These algorithmic procedures, having said that, carry out certainly not creating an exact same solution every time. Hence having a combined study of all the above mentioned algorithms, the hybrid model has been proposed in connection with multi-view similarity concept.

Table 2.6 Notations

Notation	Description
n	*Number. of e-profiles*
e_p	*e-profile of individual*
c/C_H	*e-profilesthat are clustered after similarity check*
n_H	*Number of clustered e-profiles lacking similarity check*
T_H	*Clustered set of e-profiles post validation*
D'	*Centroid reference e-profile*
ΔK_p & ΔK_q	*2 criterion functios that check similarity of pairwise profiles*
G	*Combined vector of total e-profiles*
C_v	*Centroid Vector*
GD	*Generic identifier*
H	*Health Records*
S	*Set of all health records*
D_v	*Composite Vector*

2.7 Proposed Model of E-healthcare making use of k-Centroid MVS Clustering Scheme

2.7.1 Formation of k-centroid MVS cluster

Each health document/record in the database/corpus represents an n-dimensional vector e_p, where n is the range of all the terms the record corpus has. Document vectors are made to go through some weighting plans, such as the Term Frequency-Inverse Document Frequency (TF-IDF), and normalized to have system size as unit length. The modified k-means algorithm is described as k-centroid technique in our proposed model. The concept of clustering process is to organize information (data objects) into different collections in a way that the intra-cluster similarity, in addition to the inter-cluster significant difference, is made best use of. The proposed solution indicates the fact some forms of dimension are required to determine such as similarity (i.e. resemblance) or dissimilarity.

$$Dist(e_p^1, e_p^2) = ||e_p^1 - e_p^2|| \tag{2.6}$$

There are many modern approaches for clustering that do not use any particular type of measurement, for example, probabilistic version based approach (Monegai, 2013), non-negative matrix factorization (Thilakanathan *et al.*, 2014), info logical co-clustering (Sultan, 2014), and so forth. In this chapter, however, only approaches that utilize a detailed step are stressed. Euclidean distance is perhaps one very prominent measure. It is utilized in the typical k-means algorithm. The purpose of k-means is to diminish the Euclidean distance in between clustered objects, and the clustered centroid is,

$$\min \sum_{H=1}^{k} \sum_{e_p^1 \in C_H} ||e_p^1 - C_v||^2 \tag{2.7}$$

However, cosine similarity is a more commonly used for data in a large-dimensional as well as thin space, as in .xlv or .csv type document clustering. It is a popular resemblance rating not just in text mining but also in info access. Particularly, resemblance of two file vectors e_p^1 and e_p^2 ,$Sim(e_p^1, e_p^2)$ is specified as the Cosine of the angle between them. This is equal to their inner product for unit vectors:

$$Sim(e_p^1, e_p^2) = \cos(e_p^1, e_p^2) = (e_p^1)^t e_p^2 \tag{2.8}$$

In general, the cosine measure is appealed as an alternative of k-means that is referred to as spherical k-means (Giniat, 2011). While k-means intends to decrease Euclidean distance, spherical k-means means to make the most of the cosine similarity in between .xlv or .csv type documents in a collection as well as that clustered centroid:

$$\max \sum_{H=1}^{k} \sum_{e_p^1 \in C_H} \frac{(e_p^1)^t C_v}{||C_v||} \tag{2.9}$$

The major distinction among cosine similarity and Euclidean distance is consequently the reason in between k-means and spherical k-means. The previous focus was on vector magnitudes, while the latter highlights vector instructions. The direct application in round k-means, Cosine of file vectors is likewise widely made use of in several other document clustering approaches as a core resemblance dimension. The min-max cut graph-based spectral technique is an example (Singh and Liu, 2003). In the technique of graph partitioning, document corpus is about as a chart G = (V, E) where each document is a vertex in V and each edge in E has a weight equal to the resemblance between a pair of vertices. Min-max cut formula tries to decrease the criterion feature:

$$\min \sum_{H=1}^{k} \frac{Sim(S_H, S \backslash S_H)}{Sim(S_H, S_H)} Where,\ \underset{1 \le q, H \le k}{Sim(S_q, S_H)} = \sum_{e_p^1 \in S_q, e_p^2 \in S_H} Sim(e_p^1, e_p^2) \tag{2.10}$$

The combination of cosine equation mentioned above and the min criterion equation is equal to,

$$\min \sum_{H=1}^{k} \frac{(D_v)^t D_v}{||D_v||^2} \tag{2.11}$$

There are several other methods of graph partitioning with various cutting techniques and requirement features. They are: Average Weight (Van Gorp *et al.*, 2014) and also Normalized Cut (Rocha *et al.*, 2011), each of them has been effectively requested record clustering making use of Cosine as the pairwise resemblance rating (Michalas and Komninos, 2014), (Pasupuleti *et al.*, 2016). In (Wang *et al.*, 2013*a*), an empirical study was performed to contrast a variety of standard features for file clustering. One more preferred clustering method, which is graph-based, is executed in software called CLUTO (Nguyen *et al.*, 2011). This technique's initial designs can take the document with a nearby neighbor graph and afterwards splits the chart right into collections utilizing a min-cut formula. Cosine step, the extended Jaccard coefficient can likewise be utilized in this technique to represent similarity between nearest document vectors. Provided non-unit document vectors such as -

$$x_i, y_j(e_p^1 = x_i/||x_i||,\ e_p^2 = y_j/||y_j||)$$

their extended Jaccard coefficient is:

$$Sim_{ext.Jacc}(x_i, y_j) = \frac{(x_i)^t y_j}{||x_i||^2 + ||y_j||^2 - (x_i)^t y_j} \tag{2.12}$$

Compared to Euclidean range and cosine resemblance, the extended Jaccard coefficient considers both the immensity and also the instructions of the vectors of document. This procedure has the exact same impact as cosine resemblance if the document vectors are rather represented by their equivalent system vectors.

2.7.2 *e*-health care model KMVCS

The proposed model implants a crucial attribute, known as Multi-View Similarity clustering (MVS), to create the partnership among objects which leads to gather them appropriately (Nguyen *et al.*, 2011). The dimension of partnership association is performed ahead of time. Therefore, the cluster organization is foreseen well beforehand. The MVS carries out some computation of ground-level distance to gather the items as well as allow the means to communicate with items out of the cluster. Hence, MVS is considered a technique of 3-dimensional nature that can connect information objects and can cluster them precisely. Similarity may well be found out through cosine similarity measure. In procedure of clustering, comparable e-profiles are grouped with each other. In this style, the origin factor can be specified as the patient's disease [D'].

Each of the healthcare accounts keeps collection of symptom as well as the data related to treatment of the illness of every patient. Consequently, the resemblance formula can be created as:

$$Sim(e_p^1, e_p^2) = \cos(e_p^1 - D',\ e_p^2 - D') \tag{2.13}$$

$$= \cos\left(e_p^1 - D'\right)^t, \left(e_p^2 - D'\right) \tag{2.14}$$

$$[Sim(e_p^1, e_p^2)_{e_p^1, e_p^2 \in C_H} = \frac{1}{n - n_H} \sum_{n.e_p \in\ c/C_H} Sim(e_p^1 - ne_p, e_p^2 - ne_p) \tag{2.15}$$

In the above Eqn. 2.13 Eqn. 2.14, the centroid point is stood for as D' according to the formula of both relative profiles (e_p^1, e_p^2). However, this procedure ends up being unlimited and also strenuous. An additional reference factor for the e-profile (ne_p) needs to be specified to accomplish optimum or exact similarity.

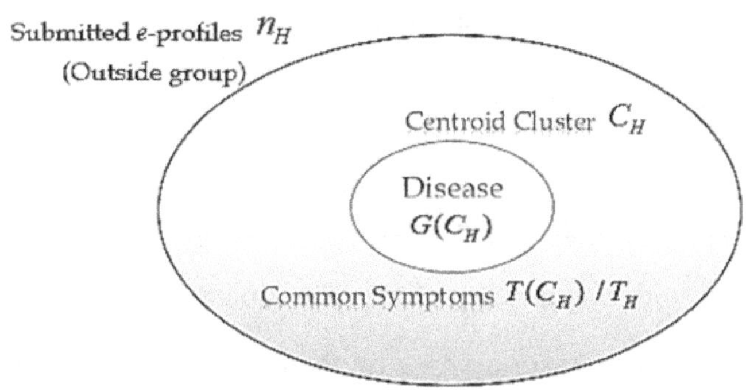

Fig. 2.4 Diseases and symptoms and their relativity of submitted e-profiles

The resemblance between two e-profiles is gauged, as well as the distinction with a third profile, which is well outside the collection, is computed. In the other formula, similarity is determined among the two e-profiles at first, and also an additional relative similarity is determined by contrasting both profiles with outside accounts such as other profiles.

For efficient personal privacy as well as safety mechanisms, security standards for data are instilled in addition to data back-up and also for retrieval of the data (de Assunção *et al.*, 2010). The secure hash formula is utilized for the purpose of privacy preservation of information sets in health domain. The nominal organizing method is embraced to bring together different e-profiles which are greatly dissimilar to each

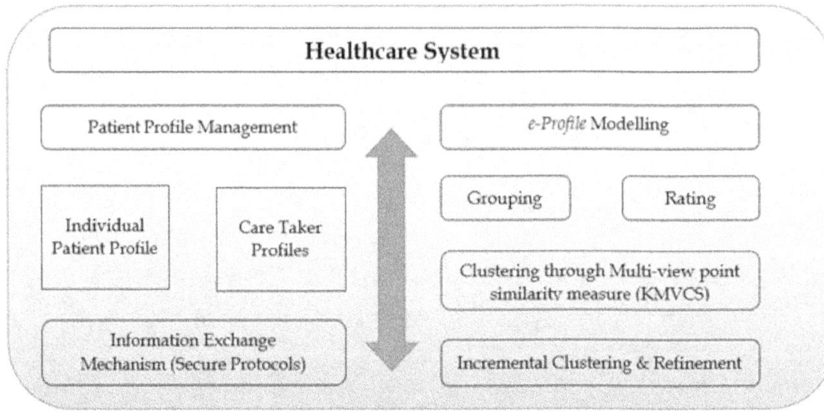

Fig. 2.5 Overall e-healthcare system model

other. The first level group includes the recognition of different accounts that applies the suitable strategy to cluster them appropriately. Ultimately, the profile-clustering makes it possible for the interaction between group participants without disclosing their personal as well as sensitive information.

2.7.2.1 Patients

Our forecast security model enables caretakers and patients to upload their profiles and to link them with one another. Each participant is considered either as a (P') the patient, a(C'), the care taker along with their typical illness and symptoms and signs are described as [Group of Patients=$\{P', C', D'\}$]. Team members are given access not only to view but also to post their message, or discuss any issue. Full access may come under one of three categories. The value of a patient for a particular illness varies in the possibility value of 0 or 1. The typical system may make use of score values and also team the people as necessary. In a similar other strategy, rankings on a range of 1-5 are given to the patients. The signs, symptoms, and other conditions are taken into consideration with the help of a method that is multidimensional. The parameters for scaling, which are stabilized may not be considered ideal for the purpose of grouping. The conditions, as well as symptoms, are ordered in nature. Grouping all the different patients into teams based upon their basic account information and wellness info is very difficult. Table 2.6 presents the percent of incidents of illness. The method makes use of multi-view ordered model for the purpose of clustering.

2.7.2.2 Caretakers

In many cases, the patients may not get involved or incorporated with e-healthcare system directly. Community assists the caretaker in the process of linking and working together with the health care system from the side of the patients. But, this sort of accounts need to be treated independently. The caretaker will be able define the total info of the person as well as launches interaction with various other community participants.

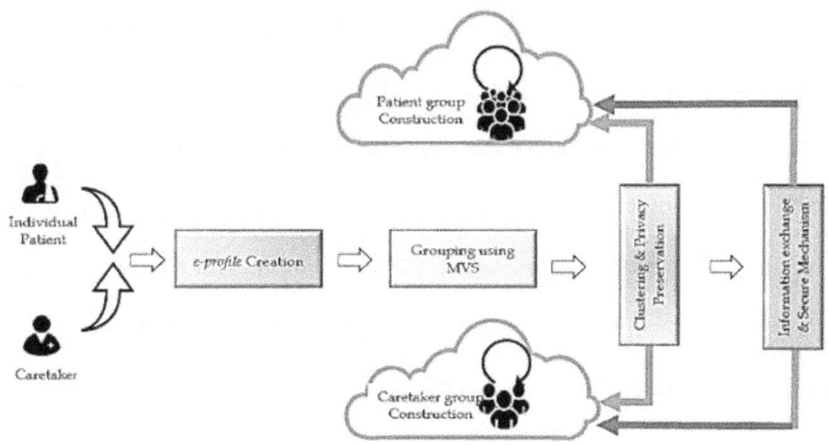

Fig. 2.6 A SaaS platform for cloud based HC-KMVCS

2.7.2.3 Collector system of healthcare domain

This system preserves user data of the individual. However, the material of every single account is hidden and also accumulated within each team. Thus, users' private info is entirely secured from various other patients (customers) inside the group. Adaptive personal privacy preserving technique is put on among profiles so that to avoid the information from being accessible to other patients within the team.

This particular system concentrates on accumulating actual accounts of both the individual as well as the caretaker. Client information is gathered with the help of the EMR information seeking form. *ElectronicMedicalRecord = Patient's personal info, Current health info of patient, Group's originator, Other group's member details,*

Secondary info. All areas or qualities of the record of EMR will be individually gathered and also categorized as an individual e-profile. Collector module applies the k-centroid MVS technique for gathering the e-profiles in accordance with their characteristics of disease.

2.7.2.4 *e*-profile of patient and creation of group

One of the first steps in this health care system is the development of e-profile. Upon entering into the system, the profiles of care taker or patient must be authenticated with the help of an authentication system. After valid authentication, the customer is required to submit the details along with the signs and symptoms as well as conditions in the EMR sheet. Every customer of an e-profile will have to produce his/her valid username along with the password for all types of data entry as well as for interaction with various other account participants. Each profile consists of private details of the patients. All e-profile users will be able to invite other patients. After the production of individual profiles, the grouping of first-level is performed. The formation of user groups is given below in algorithm 2.2.

2.8 Algorithm 2.2: Group Creation

1. Initiation: Every single patient should have his EMR linked with the e-profile
2. It is assumed that systems allow authenticated profiles only.
3. Let L be the collection of all patients in the system
4. For each patient $l \in L$; E_l is the contemplated patient team size of l
    - a. While not every patient in L is in the group
    - b. do
    - c. for every $l \in L$; who have not been associated with other group
    - d. do
    - e. l is made as host of the group in context with confidence (probability)
    - f. Prob($\mathrm{O_p}$) $= E_l/|L|$
5. if l is host, next

6. l sends invitation to similar patients to associate with group

7. end if

8. end for

9. for every l do

. a. Let O_p be the collection of l's friends who act as hosts of other clusters

. b. If $O_p \neq \phi$next

. c. l chooses arbitrarily $v \in O_p$ and associates with cluster of v.

. d. else

. e. Let J_p be collection of l's friends who associated with other clusters;

. f. if $J_p \neq \phi$ then

. g. l arbitrarily selects $v \in J_p$ and associates with the cluster of v;

. h. end if

. i. end if

10. end for

11. for every cluster of patients B do

. a. Let l be inviter of B

. b. If $P_B \geq E_l$ next

. c. B is established;

. d. end if

12. end for

13. end while

The cluster of patients (L) is built, making use of the aforementioned algorithm. The person group calls for a minimal variety of 3 entries to maintain very little personal privacy. If the person team consists of only 2 accounts, person personal privacy is merely presumed by others by carrying out the joint calculation. Team construction need to not allow any kind of hands-on entrances. Very little authentication of every single entry should be ascertained; otherwise, the profile will have to be truncated.

2.9 *k*-Centroid MVS Clustering Scheme

Disease category through signs and symptoms may not be predicted straightaway by utilizing the classical group method. Additionally, this category entails another variant such as significant disease as well as small sign or minor illness and also major sign. The category calls for precise as well as effective resemblance measurements for e-profile clustering. The advised job executes the KMVCS (Nguyen *et al.*, 2011) for clustering same type of profiles is as follows:

Step 1: A cluster consists of same type of e-profiles and also has a tendency to add an additional account and also contrasting with outside cluster.

Step 2: This step focuses on similarity that is used on a pairwise basis.

Three accounts are sent originally; the initial account is collated with the other two pairwise profiles $MVS(e_p^1, e_p^2)$ & $MVS(e_p^1, e_p^3)$. Profile e_p^2 is closer to profile e_p^1 when compared with profile e_p^3.

$$(e_p^1)^t.e_p^2 - (e_p^2)^t d_{c/C_H} > (e_p^1)^t.e_p^3 - (e_p^3)^t d_{c/C_H} \tag{2.16}$$

By applying cosine at both sides,

$$\cos(e_p^1, e_p^2) - \cos(e_p^2, d_{c/C_H})||d_{c/C_H}|| > \cos(e_p^1, e_p^3) - \cos(e_p^3, d_{c/C_H})||d_{c/C_H}|| \tag{2.17}$$

From the above formula, profile e_p^2 is comparable to e_p^3 in regards to cosine similarity (i.e. $\cos(e_p^1, e_p^2) \leq \cos(e_p^1, e_p^3)$. Profile e_p^3, which is outside the cluster, is much less similar to profile e_p^1 based upon MVS formation. The resemblance action cannot be used as is. The profile may consist of one-of-a-kind resemblances. In some instances, these resemblances results in major variant during comparison with pairwise profiles exhibiting resemblances. To solve this opposition, the credibility has to be computed for each and every comparison. The legitimacy measurement is utilized to inspect whether the similarity procedure coincides. This makes certain that any type of number of profiles that are in the corpus needs to be organized under the same cluster with an optimum similarity element.

2.9.1 Construction of MVS matrix (A) and validity score

The human illness can be at first justified based upon the incidents of signs. Each disease might contain several combinations of signs. The healthcare e-profile of each individual includes client details, a collection of symptom lattice structure, and also various other supporting info. The similarity action can be applied to each profile uploaded to the system.The MVS matrix construction is given below in algorithm 2.3.

2.9.2 Algorithm 2.3: MVS Matrix Construction (A)

1. Start
2. Let e-profile of cluster of healthcare be H
3. Let number of clusters be d.
4. for H$\leftarrow 1 : d \quad do$
5. $G_{c/C_H} \leftarrow \sum \_e_p^1 \notin C_H e_p^1$
6. $n_{c/C_H} \leftarrow |c/C_H|$
7. $end\,for$
8. $for\ p^1 \leftarrow 1 : n\ do$
9. $H \leftarrow class\,of\,e\_p\hat{1}$
10. $for\ p^2 \leftarrow 1 : n\ do$
11. $if\ e\_p\hat{2} \in C_H\ then$
12. $(AD)_{p^1,p^2} \leftarrow (e_p^1)^t.\ e\_p\hat{2}\ - (e_p^1)^t \frac{G_{c/C_H}}{n_{c/C_H}} - (e_p^2)^t \frac{G_{c/C_H}}{n_{c/C_H}} + 1$
13. $else$
14. $(AD)_{p^1,p^2} \leftarrow (e_p^1)^t.\ e\_p\hat{2}\ - (e_p^1)^t \frac{G_{c/C_H} - e_p^2}{n_{c/C_H} - 1} - (e_p^2)^t \frac{G_{c/C_H} - e_p^2}{n_{c/C_H} - 1} + 1$
15. $end\,if$
16. $end\ for$
17. $end\ for$
18. $return\ A = \{(AD)_{p^1,p^2}\}_{n\,x\,n}$
19. end

where adjacency matrix $(\mathrm{AD})_{e_p^1, e_p^2}$ is equal to $(e_p^1)^t . \, e_p^2$; supporting or auxiliary info is calculated additionally after the creation of the MVS matrix to map condition of medical diagnosis specifically. The pair-wise resemblance action contrasts the components of the initial two profiles $(e_p^1, \, e_p^2)$ and also attaches them if both profiles fall under the same class. Or else, profile e_p^2 is maintained outside e_p^1 class. This is a constant process for all gone into accounts. The validity is the measure for evaluating performance of a similarity measure and corresponding algorithm is given in algorithm 2.4.

2.9.3 Algorithm 2.4: GETVALIDITY (val, A, Percentage)

1.Start

2.for H$\leftarrow 1 : d \;\; do$

3. $\quad T_H \leftarrow [Percentage \text{ x } \mathrm{n_H}]$

4.$If \, T_H = 0 \, then$

5. $\quad T_H \leftarrow 1$

6.$end \, if$

7.$end \, for$

8.$for \;\; p^1 \leftarrow 1 : n \, do$

9. $\{Ap^1v\,[1], ..., Ap^1v\,[n]\} \leftarrow Sort\,\{Ap^11, ..., Ap^1n\}$

10.$Ap^1v\,[1] \geq Ap^1v\,[2] \geq .... \geq Ap^1v\,[n]$

11. $\quad \{v\,[1], ...v\,[n]\} \leftarrow permute\{1, ...n\}$

12. $\quad H \leftarrow class\, of\, e\_p\hat{1}$

13. $\quad val(\, e\_p\hat{1}) \leftarrow \frac{|\{ev\,[1],.........ev\,[T_H]\} \cap C_H|}{T_H}$

14.$end \, for$

15. $\quad val \leftarrow \frac{\sum_{p^1 \leftarrow 1}^{n} val(ep^1)}{n}$

16. $return \;\, val$

17. end

After the MVS matrix is created, all the entries went into the profiles are gathered as well as named separately. The legitimacy score may be applied to elucidate the closest connection between the pairs of profiles.

The initial similarity matrix is created as follows:

$$A = \{(\mathrm{AD})_{e_p^1, e_p^2}\} \tag{2.18}$$

The overall count T_H is stood for via the percent worth depending on the dimension of class H. Later on, the majority of the account data is a comparable collection of information with only minor differences. The last validity parameter of every single profile is figured out by computing the mean of all individual profiles of A. Legitimacy rating is approximated to 0 or 1. Provided any kind of profile has a legitimacy rating of 1, it represents the maximum resemblance index of grouping (Muller *et al.*, 2012).

2.9.4 Incremental clustering and refinement

The creation of the matrix of MVS, as well as the estimation of the legitimacy rating, makes sure that comparable accounts may be clustered considering the resemblance index values. Nevertheless, the condition cannot be anticipated immediately depending on a set of specified signs and symptoms. Signs and symptoms cannot be similar constantly as well as may differ time to time. Step-by-step clustering is utilized to accumulate the symptomatic details of every single profile from time to time.As new patients join the system, there should be proper allotment of patient to the concerned cluster and the concerned algorithm is given in algorithm 2.5. The refinement of clusters is being described in algorithm 2.6.

2.9.5 Algorithm 2.5: Incremental Clustering Initialization

1. Start
2. Select i seeds $C_1, C_2, \ldots\ldots. C_i$ randomly
3. For each cluster $[ep^1] \leftarrow P = \arg\max_H \{C_H^t, ep^1\}, \forall P = 1, 2, \ldots.., n;$

Table 2.7 Percentage of occurrences of symptoms of diseases

Disease	Symptom_1 (∼21%)		Symptom_2 (∼21%)		Symptom_3 (∼15%)		Symptom_4 (∼18%)		Symptom_5 (∼10%)		Symptom_6 (∼12%)	
Influenza	Fever	19%	Uncoordination	23%	Snuffle	13%	Pleuritic pain	16%	Malaise	9%	Sore throat	11%
Systemic infection	Respiratory distress	21%	Fever	24%	Tachypnoea	15%	Hypotension	18%	Decreased body weight	10%	Lethargy	12%
Hernia hiatal	Fatigability	21%	Abdominal pain	24%	Vomiting	13%	Prodrome	17%	Hyponatraemia	10%	Nausea	11%
Hyperlipidaemia	Angina pectoris	21%	Chest pain	23%	Presence of q wave	14%	Palpitation	17%	Sweat	10%	Photopsia	11%
Asthma	Cough	20%	Wheezing	22%	Chest tightness	14%	Shortness of breath	16%	Productive cough	10%	Pleuritic pain	12%
Degenerative polyarthritis	Numbness of hand	21%	Food intolerance	23%	Asthenia	15%	General discomfort	17%	Stiffness	8%	Nonproductive cough	11%
Migraine disorders	Vomiting	21%	Ambidexterity	22%	Numbness	15%	Dizziness	17%	Fever	10%	Nausea	11%
Chronic kidney failure	Orthopnoea	19%	Vomiting	24%	Oliguria	14%	Hyperkalaemia	16%	Nausea	8%	Jugular venous distention	11%
Deep vein thrombosis	Pain	19%	Swelling	23%	Shortness of breath	15%	Ecchymosis	17%	Cardiomegaly	7%	Pain in lower limb	11%
Glaucoma	Unsteady gait	19%	Distended abdomen	23%	Hyponatraemia	13%	Paraesthesia	18%	Unconscious state	7%	Agitation	12%
Dehydration	Vomiting	19%	Diarrhoea	24%	Nausea	13%	Hypotension	16%	Mental status changes	10%	Lightheadedness	12%
Pneumonia aspiration	Fever	20%	Mediastinal shift	23%	Mental status changes	13%	Clonus	18%	Unresponsiveness	7%	Decreased translucency	11%
Gastroenteritis	Breath-holding spell	21%	Abdominal pain	23%	Decreased body weight	14%	Nausea	16%	Scleral icterus	7%	Sore to touch	11%
Breast carcinoma	Paraesthesia	21%	Mass of body structure	24%	Erythema	15%	Retropulsion	16%	Burning sensation	10%	Lesion	12%
Carcinoma	Pain	20%	Mass of body structure	24%	Tumour cell invasion	15%	Lesion	17%	Decreased body weight	7%	Thicken	11%
Sickle cell anaemia	Pain	19%	Breakthrough	23%	Chest pain	15%	Shortness of breath	18%	Green sputum	8%	Abdominal pain	12%
Hypercholesterolaemia	Increased sweating	19%	Chest pain	24%	Syncope	15%	Pressure chest	18%	Shortness of breath	10%	Numbness	12%
Glaucoma	Unsteady gait	20%	Distended abdomen	22%	Hyponatraemia	14%	Paraesthesia	16%	Unconscious state	10%	Agitation	12%
Gastritis	Abdominal pain	19%	Heme positive	22%	Disequilibrium	14%	Vomiting	16%	Intoxication	9%	Nausea	12%
Hepatitis c	Distended abdomen	19%	Ascites	24%	Cough	14%	Feeling suicidal	16%	Macerated skin	8%	Ache	11%
Diabetic ketoacidosis	Vomiting	20%	Polyuria	24%	Abdominal pain	13%	Nausea	18%	Nervousness	10%	Milky	11%
Lymphoma	Welt	20%	Fever	22%	Decreased body weight	14%	Transaminitis	16%	Tinnitus	8%	Ataxia	11%
ronchitis	Wheezing	19%	Cough	24%	Chest tightness	14%	Shortness of breath	17%	Sore throat	9%	Fever	12%
Hepatitis B	Tachypnoea	19%	Inappropriate affect	22%	Projectile vomiting	15%	Yellow sputum	18%	Abdominal pain	10%	Poor feeding	11%
Endocarditis	Thicken	21%	Pleuritic pain	24%	Apyrexial	15%	Myalgia	16%	Flare	9%	Night sweat	11%

4.

$$G_H \leftarrow \sum_{ep^1 \in C_H} ep^1, n_H \leftarrow \ |C_H|, \quad \forall H = 1, ..., i;$$

5. End

2.9.6 Algorithm 2.6: Refinement

1.Start

2.Repeat

3.{v [1:n]}$\leftarrow$ $random\ permutation\ of \{1, ..., n\}$

4.$foreachep^1\_ \leftarrow 1 : n\ do$

5.$ep^2 \leftarrow v\ [p_2]$

6.$p\ \leftarrow e\_cluster\ [ep^1]$

7.Δ $K_p \leftarrow K(n_p - 1, G_p - ep^1) - K(n_p, G_p)$

8.$q \leftarrow \arg\max\{I(n_H + 1, G_H + ep^1) - K(n_H, G_H)\}$

9.Δ $K_q \leftarrow K(n_q + 1, G_p + ep^1) - K(n_q, G_q)$

10.$if\ \Delta\ K_p{+}\Delta\ K_q > 0\ \ then$

11.$Move\ \ ep^1\ to\ cluster\ q:\ \ cluster\ [ep^1] \leftarrow q$

12.$update\ G_p,\ n_p, G_q,\ n_q$

13.$endif$

14.$endfor$

15.$until$

16.$No\ move\ for\ all\ n\ profiles$

17.end

At any specific phase, upcoming signs and symptoms may result in an added risk of disease. Soon after the profiles are gathered considering the signs and symptoms, one more step is to be executed on a routine basis to merge the condition. Later on, there must be an additional chance of the occurrence of a brand-new sign that probably points to a distinct disease. This certain circumstance is termed as a clustering issue of subspace. This concern may be fixed by introducing 2 benchmark functions. The two benchmark functions ΔK_q and ΔK_p are applied to inspect pair-wise resemblance in between forthcoming accounts as well as recently upgraded

profiles. These 2 functions make sure the similarity action between the present and already existing gathered accounts by making use of inter-cluster and intra-cluster resemblance.

2.9.7 Distributed privacy preservation for clustered profiles

The k-centroid MVS clustering process computes the resemblance between health care accounts. Conversely, its effectiveness is determined based upon the patients'personal privacy variables (Xiao and Tao, 2007) In a multiparty protected interaction version, individual privacy must be safe and secure to protect an individual's sensitive info. In our suggested version, the distributed hashing strategy is executed for any kind of communication between profile participants. Admin serves as a centralized web server to keep an eye on the member's tasks. The hashed value is constructed as Sc_Cde data structure Sc_Cde (hash_key, hash_value). The scattered arbitral check is used for each and every initiation. Upon the development of each account, Sc_Cde is appended through probability of predefined nature. The e-profiles are made anonymous by incorporating a Global unique Identifier (GD). The GD is a 128-bit arbitrary number distinct to each client account, and is generated every time patients launch communication. Every time a client interacts with one more client, the Sc_Cde needs not just to be corroborated but is entirely anonymized too. By way of using this procedure, patient information cannot be recognized by various other members of the team. The distributed privacy preservation algorithm is given in algorithm 2.7.

2.9.7.1 Algorithm 2.7: Privacy Preservation $(L, C_H, P_{\text{Prob}})$

1. Initialize that Sc_Cde $\rightarrow \Phi$

. a. $L\rightarrow$ Collection of Patients in the system

2. C_H →Collection of e-profiles

3. P_{Prob} → Probability of predefined nature of of C_H

4. R_{Prob} → Probability at Random

5. Let the server choose arbitrarily $l \in L$

6. While not every profile is in C_H are in Sc_Cde

. a. if l receives Sc_Cde at first instance

. b. then

. c. l at random generates $0 \leq R_{\text{Prob}} \leq 1$

. d. if $R_{\text{Prob}} \leq P_{\text{Prob}}$ then

. e. l generates at random GD(l) where GD stands for a global unique identifier

. f. CH(l)=CH-CH(Sc_Cde);

. g. Sc_Cde.Assign (GD(l),CH(l));

7. end if

8. end if

9. if l associates to other profile L' from the corresponding group at random,

. a. L' € R(p)

10. Then

11. l has to send Sc_Cde to L' for random check

. a. l = l';

12. while Sc_Cde is sent to administrator do

13. l sends Sc_Cde to administrator with the chance of P_{Prob} for

14. if Sc_Cde is not sent to administrator

15. Then

16. l runs one more random walk

17. end if

18. end while

2.10 Experimental Results and Analysis

To show the functionality of the anticipated model, the privacy preservation of the medical care community is evaluated in numerous varied circumstances. The total cross-breed design is designed as well as created in the C++ programming language and also tested on the GreenCloud simulation environment. This simulation software application is set up in the Ubuntu 14.04.2 LTS OS with an arrangement of 2.8-GHz Intel Dual Core CPU and 8-GB RAM. The GreenCloud simulator is much more efficient on many network equipment's with different switches such as Core, Aggregations, as well as Access switches. These switches are in charge of linking to different network makers with various bandwidths.

The industrialized prototype version is carried out, as well as a comprehensive assessment of the Symptom-X data set is performed (Sultan, 2014) It is a comprehensive analysis and ends with the generation of a table of symptoms as well as associated illness. The KMVCS emphasises high quality, performance, as well as personal privacy conservation for the healthcare area system. We can say that Multi-View similarity measure(MVS) is a robust measure and as it is considering all the views from profiles outside the cluster it achieved best results compared to other similarity measures. The evaluation of KMVCS performance concentrates on FScore as well as accuracy with all group occasions. FScore is an equally heavy combination of accuracy for all group occasions (Wu, 2012) These two parameters are exhibited as adheres to:

$$Accuracy = \frac{|C_H(e_p) \cap C'_H(e'_p)|}{C'_H(e'_p)} \tag{2.19}$$

$$Group\, event = \frac{|C_H(e_p) \cap C'_H(e'_p)|}{C_H(e_p)} \tag{2.20}$$

$$FScore = \sum_{i=1}^{e_p} n_H/n \max_j \ (F_{i,j}) \tag{2.21}$$

Where,

Table 2.8 Results of the accuracy of clustering

Disease/Schemes	**KNN**	**KMVCS (ΔKp and ΔKq)**	**SVD**	***k-means***
Gastritis	0.5	0.5	0.6	0.1
Influenza	0.1	0.9	0.9	0.2
Epilepsy	0.3	0.9	0.8	0.1
Hypercholesterolaemia	0.4	0.4	0.2	0.2
Paranoia	0.7	0.1	0.8	0.5
Asthma	0.5	0.8	0.5	0.4
Neoplasm	0.4	0.8	0.5	0.6
Hyperlipidaemia	0.6	0.8	0.8	0.4
Glaucoma	0.1	0.6	0.3	0.2
Kidney disease	0.4	0.8	0.4	0.2
Morbid obesity	0.8	0.6	0.5	0.8
Pneumonia aspiration	0.1	0.2	0.7	0.6
Hepatitis B	0.4	0.6	0.3	0.3
Bronchitis	0.2	0.1	0.1	0.4
Diabetes	0.4	0.4	0.2	0.5

$$\mathrm{F}_{i,j} = \frac{2 \times P_{i,j} \times Q_{i,j}}{P_{i,j} + Q_{i,j}}$$

$$P_{i,j} = \frac{n_{i,j}}{n_j},$$

$$Q_{i,j} = \frac{n_{i,j}}{n_i}$$

where n_i represents the number of signs mapped to disease i, n_j represents the variety of e-profiles designated to cluster j, and $n_{i,j}$ also signifies the number of e-profiles shared by i as well as j. In the substitute atmosphere, greater than 5,000 e-profiles are created by incorporating both client as well as caretaker accounts. After official collection, customers are permitted to interact with other users within the collection.

Table 2.9 Results of *FScore* for clustering

Disease/Schemes	KNN	KMVCS (ΔKp and ΔKq)	SVD	*k-means*
Gastritis	**0.831**	0.351	0.724	0.346
Influenza	0.932	**0.868**	0.366	0.782
Epilepsy	0.504	0.657	0.159	0.416
Hypercholesterolaemia	0.013	0.225	0.934	**0.801**
Paranoia	0.33	**0.85**	0.473	0.644
Asthma	0.319	**0.991**	0.452	0.627
Neoplasm	0.523	0.769	**0.849**	0.651
Hyperlipidaemia	0.464	0.34	0.205	0.474
Glaucoma	**0.983**	**0.901**	0.161	0.192
Kidney disease	0.79	0.523	0.658	0.297
Morbid obesity	0.762	**0.917**	0.01	0.394
Hepatitis B	0.218	**0.862**	**0.861**	0.747
Pneumonia aspiration	0.378	0.696	0.381	0.574
Bronchitis	0.227	**0.865**	0.403	0.478
Diabetes	0.073	0.149	0.414	0.467

The interactions are revealed as various interesting blog posts, comments, and also recommendations. Every day, roughly 2,000 brand-new articles and also greater than 3,000 reviews are shown for 150 clusters. The varieties of profiles, articles, and also checks out might differ for each cluster. The entire evaluation is duplicated 100 times by using the same information collections. For all three metrics, a higher worth stands for extra accurate clustering as well as more stringent personal privacy preservation.

2.10.1 Accuracy

The precision of four clustering algorithms on 20 information collections is graphed in Fig.2.7. One more technique for determining clustering is FScore tabulated in Table 2.8. Two requirement functions are chosen to perform step-by-step clustering. Nonetheless, in the speculative examination, these two criteria are integrated. The efficiency of this recommended model is modest and also substantially consistent in contrast with that said of other similar algorithms. The performance of the model is also constant with that of the SVD-based distributed clustering formula. Application of the version to a static information set yields superior results. A t-test (Wu, 2012) is executed with a 5 $\%$

Table 2.10 Overall latency comparison of KMVCS (Δ Kp and Δ Kq),SVD, KNN, k-means

List of Schemes	Accuracy									
	***k* = 5**	***k* = 10**	***k* = 15**	***k* = 20**	***k* = 25**	***k* = 30**	***k* = 35**	***k* = 40**	***k* = 45**	***k* = 50**
KMVCS (Δ Kp *and* **Δ Kq)**	0.086	0.802	0.442	0.87	0.094	0.856	0.258	0.923	0.901	0.865
SVD	0.445	0.657	0.069	0.767	0.727	0.55	0.455	0.585	0.704	0.503
KNN	0.019	0.718	0.346	0.674	0.276	0.305	0.798	0.261	0.58	0.796
k-means	0.902	0.063	0.964	0.247	0.974	0.113	0.843	0.248	0.147	0.18

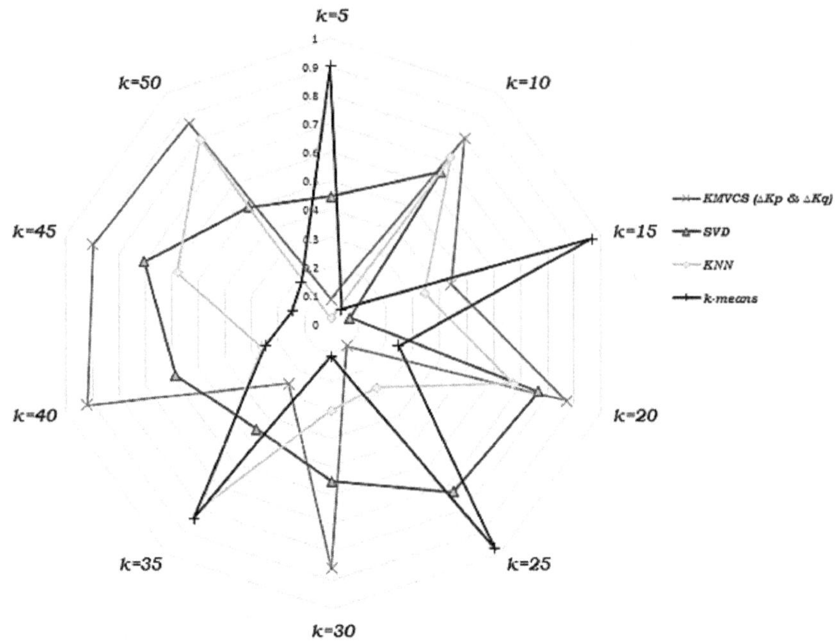

Fig. 2.7 Accuracy results of 20 data sets (test case k = 5)

importance degree. The 2nd requirement functions are combined to put together a null hypothesis with a mean of 0 (Tables 2.7 as well as 2.8). Nonetheless, combining these 2 functions did not lead to inconsistency in the results.

2.10.2 Latency

The group size may differ among different collections. However, the latency of the system can not be straight reviewed. A bigger group dimension may hamper mutual processing as a result of a lot more gathering.Thus, each cluster dimension is limited to 100 to compute the client-side latency (Fig.2.8). Determining the overall latency of the system is extremely challenging. In the examination, the basic deviation value of k = 10 is made use of, and the cluster size is limited to k = 100. For large-scale peer-to-peer communication, O(klog n) is repaired for the client-side computation. For this reason, the efficiency of the system is higher after fine-tuning the k-means algorithm with multi-view point similarity in addition to personal privacy preservation. The KMVCS is confirmed to perform sensible clustering and also safety and security specifications.

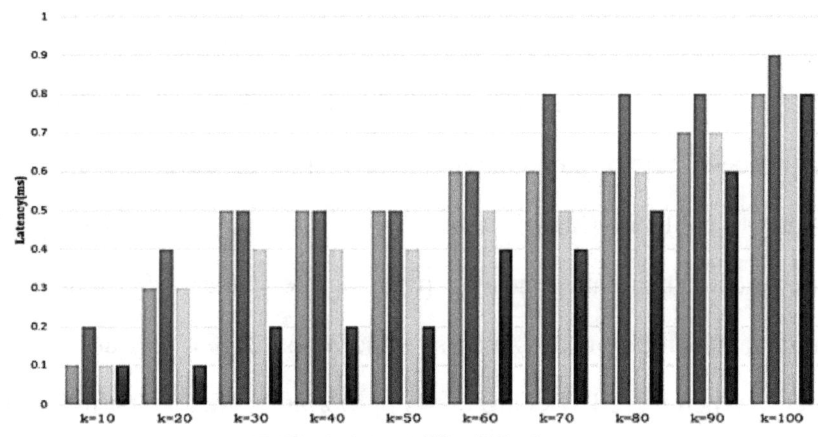

Fig. 2.8 Overall latency comparison of KMVCS (ΔKp and ΔKq),SVD, KNN, k-means

CHAPTER 3

HEALTHCARE DATA PRIVACY AND PRESERVATION IN THE CLOUD USING OPTIMAL ITEM SET MINING

3.1 Introduction

In the medical industry, Electronic Health Records (EHRs) have indeed been routinely implemented in a large number of healthcare facilities in an endeavour to improve the excellence of care of patients while also increasing efficiency and profitability of provision of healthcare (Sharifi *et al.*, 2013), (Qamar *et al.*, 2016). Typically, EHRs are generated and maintained inside a particular healthcare office, such as a hospital, clinic centre, or doctor's office (Andersen and Saus, 2017), and are not shared with other offices. When these EHRs are used appropriately, they assist in various healthcare activities as well as correctly identifying illnesses. It is apparent from the customers' point of view that a wide variety of benefits exist for clients participating in e-Health systems (Yang *et al.*, 2015). In general, individuals need different care services, such as hospitals, physicians, or nursing homes, throughout their lifetime. Given the fact that a patient's medical records are available online and can be distributed between various care providers (Taneja *et al.*, 2015), this could facilitate the patients in conveying information about previous conversations as well as a diagnosis. It also assists the healthcare professionals in making better decisions through the use of the exhaustive nature of consistent electronic archives (Majeed, 2019).

Despite the fact that they are necessary and provide lucrative opportunities to defend the previously described vulnerabilities (Wang *et al.*, 2015), the investigation here has not yet addressed all of the additional security vulnerabilities (Meingast *et al.*, 2006).

Due to the fact that these systems are responsible for monitoring and supervising health-related information, information assurance and access control are critical (Vora *et al.*, 2018). In the same way, in order to provide assistance to the elderly or ill, it is necessary to ensure collaboration with the caregivers (Dankar and Al Ali, 2015). From one point of view, this suggests that the system will also take care of caregivers' near to personal information, such as their contact information (Henze *et al.*, 2016). Then, knowing a patient's caretaker might lead to the discovery of sensitive medical information about the patient (Haldorai and Kandaswamy, 2019). For instance, recognizing which entity is dealing with patients may make it possible to deduce the nature of his medical problems. As a result, all security concerns need be addressed in order for these systems to be widely received and transmitted (Huang *et al.*, 2009). This problem is addressed by a set of security-preserving standards that are based on a new e-Health platform which might safeguard the data of its customers while also ensuring that all of the critical data is accessible to appropriate care provisioning components (Loukides *et al.*, 2014). To put it another way, the assurance of safeguarding would not compel restrictions on the number of bureaucracies or the sort of care provided by the system.

To handle these issues, we provide a healthcare data security strategy that uses item-set mining that is independent of a hierarchical organizational structure (Sahi *et al.*, 2017). Our goal is to create a system that allows all offices (or geographical locations) to benefit from their combined expertise, without gathering patient-level information. Data security is concluded with the grouping and itemset mining in the evaluation. This itemset mining aids in the identification of the best characteristics for the parameters, with the purpose of reducing risk and increasing information usefulness in order to reduce costs and increase efficiency. The dissemination of health records cannot be prevented; however, the access to the data that is being shared may be prevented with the aid of newly developed technology.

The Chapter is organized in the following manner: Section 3.2 presents the results of a survey on the protection of personal information in e-Health care data. Section 3.3 highlights the factors that lead to the need for e-Health care security and privacy;

Section 3.7 includes a full overview of the approach, as well as the approaches that have been suggested. Section 3.10 describes the findings of the intended research, and Section 3.11 presents the conclusion of the research and Section 3.12 projects future directions for the research.

3.2 Literature Survey

An investigation of the use of privacy-preserving methods and their application to e-Health care data was conducted by GardiyawasamPussewalage, H. S., and colleagues (Pussewalage and Oleshchuk, 2016). While studying the privacy needs and obtaining high security, it is important to determine whether or not the offered techniques are consistent with the needs that have been defined. Finally, the author discussed several potential future paths for safeguarding patient health information, which will undoubtedly result in more efficient and reliable health solutions in the future.

A privacy preservation approach known as Privacy Preserving Ridge Regression (PPRR) was developed and implemented by (Chen *et al.*, 2018). for use with high-dimensional data. The protocol under consideration has a system in which every client encrypts its information and delivers it to the individual involved. They keep a straightforward model of each and every client's information without taking into consideration their consent. It is necessary to utilize an encryption approach that has special homomorphic features in order to carry out PPRR. This approach is effective for handling large amounts of data across numerous clients using PPRR.

A private cloud, as suggested by (Nadaf and Patil, 2016) in 2016, would allow for the integration of security into an e-Healthcare system. It is possible to ensure the security of stored health data by encrypting the information before saving it on a cloud server. The features of the system that stand out include efficient administrative tools, the security of data storage, and data recovery. Before putting the health information on a cloud server, it is encoded to ensure that it can be accessed securely on dispersed storage.

C. Sun and colleagues (Sun *et al.*, 2014) introduced a customized security problem in 2014, in which different characteristics may need the purchase of different levels of insurance protection. In order to deal with this problem, the privacy mechanism is combined with a randomized answer. This approach gained maximum precision in frequent mining than the standard ways, based on its measuring of similarity, by providing varied protection levels for different features.

In (Nayahi and Kavitha, 2017), they investigated an anonymization technique that was based on process of clustering and was adaptable to both similarity attacks and probabilistic inference attacks, among other things. Hadoop distributed file system (HDFS) distributes anonymized information. The technique achieves a better exchange between protection and usefulness. As for several classifiers, the author assessed data usefulness, accuracy, and F-message. Analyzed indicators of performance include Accuracy: The data indices of the suggested clustering algorithms are higher than the existing methods, and they save the processing time of the offered method protection.

In 2017,(Rahman *et al.*, 2017) introduced a model (PriSens-HSAC) that is intended to overcome the two security vulnerabilities previously described. As far as the present study is concerned, it is the prime foundation for extending safety in healthcare systems that use RFID and access control strategies in the field of Radio Frequency Identification (RFID). The suggested technique tested the execution and precision of the framework by putting it through its paces in several real-world scenarios involving diverse client roles such as physician, emergency care provider, and pharmacologist.

Several research publications on privacy-preserving besides disaster recovery in the e-Health cloud were addressed by (Sahi *et al.*, 2016) in 2016 as part of a panel discussion. The disaster recovery proposal was developed, which includes two approaches, the Security-Preserving Approach (SPA) and the Privacy-Preserving Approach (PPA). E-Health Records are protected by the Security and Integrity Safety Act (SPA), and Personal Health Records are protected by the Personal Health Records Act (PPA), which is a competent verification strategy that assures the safeguarding Individual Health Records. It is already discussed how coordinated approaches and the disaster recov-

Table 3.1 Privacy Preservation –A Comparative analysis

S.No	Name of Technique	Pros	Cons
(Nadaf and Patil, 2016)	Privacy Preserving Secure Data Storage and Retrieval	▪ Uses Advanced Encryption Standard(AES) ▪ Robust against hacking	▪ Uses simple algebraic structure ▪ Hard to implement with software
(Nayahi and Kavitha, 2017)	Anonymization Based Privacy and Utility Preserving System	▪ Applies anonymization to prevent similarity and probabilistic attack ▪ Better trade-off between privacy and utility	▪ No sensitive item set mining explored
(Rahman et al., 2017)	Privacy Preservation for RFID based health care systems	▪ Applies privacy preserving authentication protocol ▪ Solves tradeoff between privacy and scalability	▪ Traditional anonymization techniques are not applicable to this domain
(Chen et al., 2018)	Privacy Preserving Ridge Regression	▪ Uses linear regression model ▪ Works well for high dimensional data	▪ Applies limited form of homomorphis m

ery plan may ensure cloud companies' dependability and security (Sahi *et al.*, 2016, Harikumar and Surya, 2015, Alam *et al.*, 2014, Sharawi *et al.*, 2017, Rajest and Suresh, 2018).

3.3 Background

The Patients' medical records may need to be obtained from a variety of sources, each with a different policy on the release of health information. This might be necessary for health-care-related and well being research projects. Because of the substandard quality and sensitive nature of health care information, the privacy of patients' medical information is a significant issue when their health care data is used for research purposes. Without personal data, the correlation and the incorporation of health data into the privacy of each patient's premise is a challenge. Following that, data clustering and subsequent information security of medical data are discussed extensively in this investigation (Rajest and Suresh, 2018, Adanov *et al.*, 2019, Sandanasamy *et al.*, 2014, Rao and Akiri, 2014, Kumar, 2014, Reddy and d'Almaine, 2015, Khanna *et al.*, 2015, Islam *et al.*, 2015).

The Chapter discusses the subject of protecting the privacy of electronic health records and enhancing the level of protection of a patient's information kept in the cloud. In the proposed study, two separate medical datasets from the University of California, Irvine (UCI) repository of machine learning are taken into consideration. As a first step, the K-medoid clustering method is used to determine the differences between the illnesses. This is accomplished by generating distance vectors among two intervals of data points. The clustered health care data is protected using an innovative approach known as optimum itemset mining (PMUIM), in which the ideal threshold value is determined using a meta-heuristic algorithm known as Modified Whale Optimization (MWO). PMUIM is a technology that was developed by the University of California at Berkeley. Fig.3.1 depicts a diagrammatic illustration of the privacy-preserving aspects of electronic health care data processing.

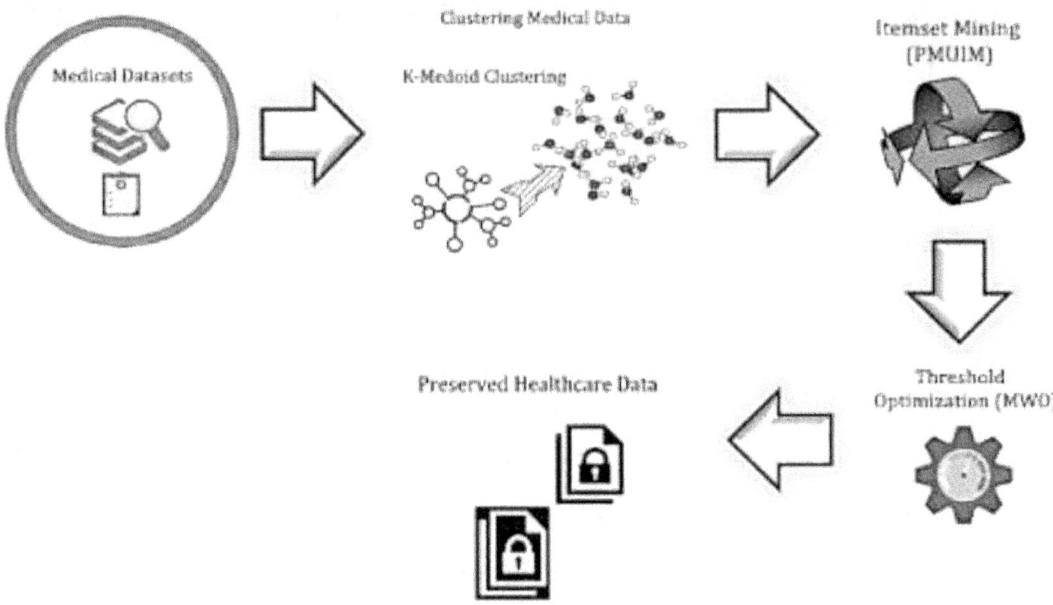

Fig. 3.1 *e*-Health care data security is diagrammatically shown in this illustration

3.4 *k*-Medoids Clustering

In comparison to K-means (Harikumar and Surya, 2015), K-medoids method is a clustering approach that's much more resistant to outliers. K-medoids uses a clustering method to solve a predefined problem, which implies the output of the clustering method is also intended to serve as a solution to the problem. This K-medoids technique focuses on selecting genuine data points as templates, is less susceptible to noise and anomalies, and provides a more robust result.

It is the absolute error criteria, rather than the Sum of Squared Error (SSE), that the K-medoids approach seeks to reduce in its execution. This procedure works much like the K-means clustering technique; it first iterates until the characteristic item is found, then checks to see whether it's the medoid of the cluster. The Method contains the fundamental steps of K-medoids clustering algorithm.

Instead of using random points for m, the K-medoids clustering algorithm utilizes certain circumstances when a random point xi replaces a representative point m. During this stage, the membership of the points that were initially assigned to m is verified to see whether there has been a change in membership. One of the two possible changes to these points' membership occurs in either one of the following ways.

The *xi* (the new representative point) or any of the previous sets of representative points are now more accessible, and these locations may now be found closer to each other. It is possible to compute the cost of switching by using the absolute error criteria for K-medoids. To estimate the total cost of re-assignment, we need to first estimate the cost of exchanging. The k-medoids clustering algorithm is given in algorithm 3.1 given below.

3.4.1 Algorithm 3.1: *k*-Medoids Clustering

1. Choose *K* points to serve as the first representative objects.
2. Continue the cycle
3. Designate every other point to the cluster that contains the item that is closest in size to it.
4. Choose a non-representative item *xi* at random from the list.
5. Calculate the total cost S of switching the representative item m with the corresponding item *xi*.
6. If S is less than zero, swap *m* with *xi* to produce a new set of K representative items in the original set.
7. till the convergence criteria is satisfied.

The amendment of the K-medoids classification termed Partitioning Around Medoids (PAM) method is presented in order to address the issue of performing many swap operations while obtaining the final representative points for every cluster (Harikumar and Surya, 2015). This method is a dissimilarity matrix algorithm, operating on a dataset's dissimilarity matrix. Until it reaches convergence, PAM minimizes the optimal solution by exchanging all of the non-medoid points with all of the medoids repeatedly.

However, although K-Medoids is more resilient than K-means, the computational cost of K-Medoids is larger, making it unsuitable for big datasets. PAM was also used in conjunction with a sampling approach to develop the Clustering LARge Application

(CLARA) algorithm, which was later adopted. CLARA evaluates a wide range of samples and implements PAM to each one of them in order to get back the set of optimum medoids at the end of the process.

3.5 Optimal itemset mining - Preserving Maximum Utility Itemset Mining

The Preserving Maximum Utility Itemset Mining has been regarded as a key topic in recent years (PMUIM). An itemset mining process may be useful for indicating items which utilize profit while considering quality, rather than using a frequent itemset (FIM) or perhaps an association rule approach. There are various high utility itemset mining (HUIs) methods developed to handle the exponential search space required to identify the Preserving Maximum Utility Itemset Mining (PMUIM) when the number of possible items and the repository was quite huge.

3.6 Whale Optimization Algorithm

The fruit fly optimization algorithm (FOA) initially suggested by (Pan, 2012) in 2012, who supplied an effective and also very easy technique to deal with the complex optimization complications, replicates the smart foraging behavior of fruit flies or even vinegar flies in food discovery. As a pretty new marketing strategy motivated by swarm intellect, the artificial honey bee colony protocol (ABC) proposed by Karaboga (Pan, 2012) in 2005 copies the foraging actions of honeybees, which is composed of 3 types of bees: employed, onlooker, as well as a precursor or scout honey bees. Since 2005, researchers have devoted themselves to the hunt techniques and also applications of ABC (Karaboga, 2005).

In 2016, SeyedaliMirjalili, as well as Andrew Lewis, initially proposed a new meta-heuristic optimization procedure (specifically, Whale Optimization Algorithm, WOA) copying the seeking habits of humpback whales (Sharawi *et al.*, 2017). The following is actually the awareness of whales in brief. Whales are considered the most significant creatures worldwide. An adult whale may grow up to 30m 180t and lengthy weight. Whales are actually mostly considered as killers, which certainly never sleep because they must breathe coming from the surface of oceans, and also fifty percent of whose intellect only sleeps.

As per study, whales possess usual cells in particular places of their brains comparable to those of humans called pin mobiles. These cells are responsible for judgment, feelings, and social actions in humans. It has been verified that whales can easily think, find out, court, interact, and come to be emotional as an individual carries out, yet definitely along with a much lesser amount of brilliance. Fig. 3.2 reveals their particular looking procedure of humpback whales. This scouring action is actually named the bubble-net feeding strategy (Sharawi *et al.*, 2017) Humpback whales choose to hunt aggregation of krill or even little fishes near to the surface, whose foraging is actually performed through creating unique blisters along a group or '9'-shaped path as received Fig. 3.2.

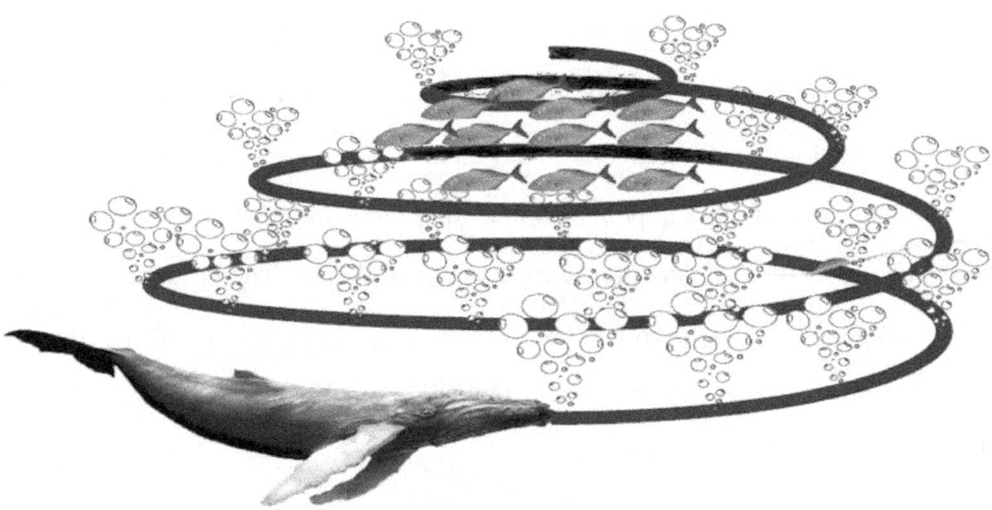

Fig. 3.2 Humpback whales' bubble-net eating habits

3.7 The proposed K-Medoid Clustering Model

The k-medoids problem is a clustering problem equivalent to k-means. Both the algorithms, such as k-medoids and k-means, divide up the datasets (breaking them into groups) and then attempt to reduce the variability among points identified as being in a cluster as well as the variance between those points that are designated as the cluster's center. In comparison to the k-means process, the k-medoids picks real information points as the centre (medoids or prototypes) and hence permits greater interpretability of the cluster centers than other one, where the center of a cluster is not always among the input data points, which is typical among points in the defined cluster). K-medoids could be used with approximate dissimilarity procedures, although k-means normally require Euclidean distance for effective services. It is more resistant to outliers and noise than k-means because k-medoids diminishes an amount of pair-wise considerable distinctions instead of an amount of squared Euclidean ranges.

Additionally, the partitions (separated medoids) and K-means partitions (aggregated medoids) are somewhat different from each other. Both algorithms make attempts to reduce the squared error; however, the K-medoids approach is more resistant against faults than the K-means approach. For instance, the K-means algorithm advises selecting the centroids, while the K-medoids takes care of gathering information or data points. A medoid may be supplied as the cluster's object, which then averages the variance over all of the supplied items. The difference between k-medoids and k-means is related to the difference between mean and median, wherein mean denotes the average value of all data points combined, and median denotes the value towards which the products of all data points are evenly distributed. The fundamental concept behind this approach is to compute the number of K agent objects, also known as medoids. Following that, the set of result medoidsis created, with each item in the data set being assigned to the medoid next to it.

3.7.1 Steps of the K-Medoid Clustering

3.7.1.1 Initialization

The K-medoids clustering algorithm is initiated through selecting the arbitrary random point k for the specific cases. This value is replaced by representative point n. The membership of point is represented via value m.

3.7.1.2 Medoids

Selection: The overall medoid value is selected by examining the distance amongst every 2 data points of all considered objects. Here, the Euclidean distance procedure is thought about for range computation. Subsequently, the above points be now nearer to k (new representative point) or can be closer to any of the other set of representative points. The cost of exchange is determined as the absolute error requirement for K-medoids. Each relocation operation, the cost of exchange is determined, and this adds to the total cost function. The elaborated k-medoid cluster formation algorithm is given in algorithm 3.2.

3.7.1.3 Algorithm 3.2: Elaborated K-medoid Cluster Formation Algorithm

1. Initialization:

. a. Define the number of cluster *k*

2. Select:

. a. representative objects through *k* points, the *k* clusters are formed through *k*-medoids *m*.

3. Repeat:

. a. Allocate every single point to the cluster with the nearest medoid *N*.

. b. Arbitrarily chose a non-representative object *ni*

. c. Calculate the over-all cost of swapping S_w, through *m* with *ni*

4. If $S_w < 0$:

. a. Exchange N with *ni* to acquire new medoid sets.

. b. End when merging benchmarks are met.

Nevertheless, as per our necessity here, the Euclidean distance measure is deliberated for distance computation in Eqn. 3.1.

$$Calculated\,Distance = \sum_{i=1}^{n} (A_i - B_i)^2 \quad (3.1)$$

Table 3.2 Description of parameters for distance calculation

Parameter	Description
N	Total number of data points
I	Total number of iterations
Ai	Distance measure of initial data point of ith object
Bi	Distance measure of subsequent data point of ith object

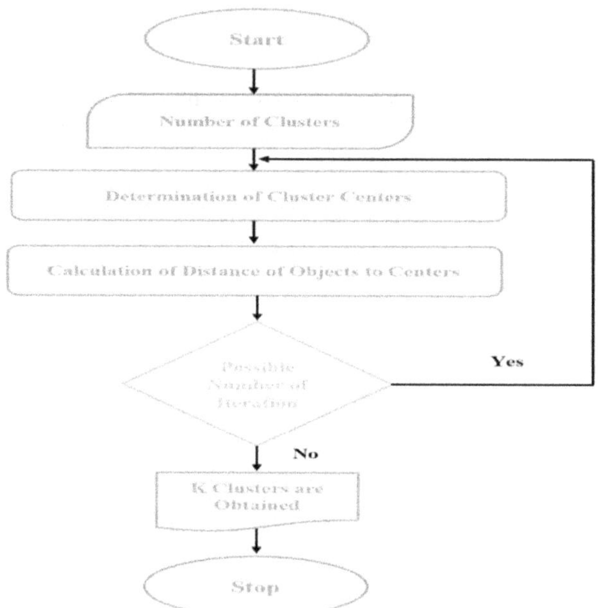

Fig. 3.3 Flowchart of K-Medoid Clustering Model

3.7.1.4 Construction of Cluster

The preliminary cluster is made by allocating each object to the nearest medoid value.

3.7.1.5 Update the Designated Medoids

The job of computing new medoid in every single cluster is that it diminishes the total distance among objects in the described cluster. If in case the newly formed medoid is contented one, then update present medoid value with the existing cluster.

3.7.1.6 Assignment of objects to every single Medoid

Through assigning each data point to the closest medoid, the grouping end result will certainly be obtained. Review the complete amount of the proximity from all objects i.e., initialized n objects to their medoids. Then, cancel the completion of the clustering algorithm if the calculated total is actually identical towards the previous one.

By using the K-medoid protocol, a patient's health care information is actually clustered according to their health condition. This are going to assist to focus on the client's illness based upon the complications to make sure that they will definitely obtain helpful treatment in a correct method. Coming from the clustered information, the client's info is gotten by utilizing the itemset exploration. In the itemset mining, the ideal limit value is actually decided on based upon the proposed optimization algorithm referred to as MWO.

3.8 Privacy Preservation approach for Medical Dataset

The mining treatment has been applied for the skilled review of a patient's clinical information to predict the illness. This investigation may discover regular health conditions from the taken into consideration UCI medical dataset with the help of the itemset mining strategy.

Acquiring information about the frequently taking place health conditions isn't a basic process. The data compilation pertaining to these kinds of diseases is achievable with itemset mining. Below, the person's health care reports are actually extracted by the PMUIM protocol.

3.8.1 The Preserving Maximum Utility Itemset Mining

The primary objective of the PMUIM protocol is actually to reduce the utility value of each item prepared i.e., health care records, by updating the utility values of items consisted of in the UCI datasets. The PMUIM protocol measures for mining the frequent itemset are actually defined below in algorithm 3.3.

3.8.1.1 Algorithm 3.3: PMUIM (*D,SI,T*)

Input - D : Transactional Database, SI: Set of Sensitive Items, T: Initial Minimum Utility Threshold

Output- Sanitized Database and high utility item sets.

1. Begin

2. Initialize D=Medical Database, T=Initial minimum utility threshold.

3. Repeat

4. Initialize T value in MWO to a value between 0 to 1.

5. Repeat

6. $diff$=the difference between the utility of sensitive high utility item set and extra amount above the minimum utility threshold.

7. M=Project the database of transactions that are having high utility sensitive item set.

8. Repeat

9. if $diff > 0$

10. Modify the quantities of high utility items in those transactions by subtracting required amount in them.

11. else

12. Continue.

13. Until all transactions in M are over.

14. Until every sensitive item set

15. Until accuracy reaches high.

16. end PMUIM.

The utility value is intended in the sensitive data items and also receipts the maximum sensitive data in transaction process. The threshold value varieties from 50 to 1500, and for accomplishing high accuracy, the optimal threshold value is originated by means of the MWO algorithm.

3.9 Threshold Optimization

The modified frequency values of items have aided mine the utility value itemsets. Additionally, depending on the maximum utility threshold value, the frequency values of items included in the medical datasets. The threshold value is in the range of 100 to 2000, and the appropriate threshold is determined by applying the MWO algorithm to a set of data.

3.9.1 Algorithm of Whale Optimization

It was developed by the authors Mirjalili and Lewis and has been named Whale Optimization (Sharawi *et al.*, 2017). It was influenced by the hunting attitude of humpback whales, although it is responsible for enhancing the threshold value of PMUIM operation.

3.9.2 Proposed Modified Whale Optimization Algorithm

Because of the velocity update of the Particle Swarm Optimization (PSO) method, this model may adjust random value selection, as shown in the Eqn. 3.9, and those stages are detailed in more detail further down the continuation.

3.9.2.1 Steps Involved in MWO Algorithm

3.9.2.2 *Initialization*

As a starting point, the random threshold values are selected as well as all of the search agent parameters that are initialised, and the population size is represented by counting the list of elements in the population.

3.9.2.3 *Fitness Computation*

The suitability is represented as fitness computation of the proposed job is assessed following the optimum threshold selection specified in Eqn. 3.2.

$$Fitness\,Function = Optimal(\tau) \tag{3.2}$$

Table 3.3 Description of parameters for MWO

Symbol	Meaning	Symbol	Meaning
T	the present iteration	*	the position vector
$\vec{H}$	Primary coefficient vectors	$\vec{C}$	a position vector representing the finest arrangement
$\vec{y}$	Second coefficient vectors.	.	a multiplication carried on element by element
$\vec{C}$	the vector indicating the location of the best solution found so far		

3.9.2.4 Updating Process of MWO

A. Surrounding the Prey : This is the exact moment when the humpback whale encloses its prey (small fish), updates its position in the direction of the optimal arrangements through the growing number of repeats, which is defined by,

$$E = |\vec{y}.\bar{C}^{*}(t) - \bar{C}(t)|, C(t+1) = C^{best}(t) - H.y\vec{1} \tag{3.3}$$

where the location or position on a vector of the best prearrangement is represented by $\vec{C}$ and represents the position vector; the current iteration is defined via t. The $\vec{H}\,and\,\bar{y}$ is represented as coefficient vectors. $(\cdot)$ is the multiplication and modification of element components by other element components (mod function). Finally, the coefficient values are considered by,

$$y\bar{1} = 2y\bar{2}\,.\,\vec{r} - y\bar{2} \quad and\, y = \bar{2}\,.\,r \tag{3.4}$$

After each repetition, the value changes only linearly between 2 and 0, and r indicating that the value is random (0, 1).

B. Bubble net attacking Model : Two different methodologies have been implemented for fresh CH (vehicle clustering) updates.

Shrinking encircling mechanism: The number was manipulated so that it decreased as and it ranged from 2 to 0, using the Random value of vector (-1 to 1).

Spiral updating position: Between the location of prey and whale, a spiraling condition is constructed to replicate the developments in helix shape of humpback whales as required.

$$\vec{C}(t+1) = \vec{H}.e^{bt}.\cos(2\Pi s) + \vec{C}(t) \tag{3.5}$$

Equation 3.5 describes the extraction process of $i\,th$ prey and b predators as a constant condition of affairs spiral with a logarithmic scale. Let us imagine that the outside system, as well as the spiral modeling, is to be renegotiated as well. Since there is a 50% potential of the values for both contracting and the spiral model both being more than or equal to the current situation, let us model so and pretend that this is possible by,

$$\vec{C}(t+1) = \begin{cases} \vec{C}(t) - y\vec{1}.\vec{H} & if\ i < 0.5 \\ \vec{H}'.e^{bs}.\cos(2\pi s) + \vec{C}(t) & if\, i \geq 0.5 \end{cases} \tag{3.6}$$

Based on the above equation, the letter s represents arbitrary value between -1 to 1, the constant *b* and *i* is ranging from ([0, 1]). In order to demonstrate this modern behaviour, we anticipate that there will be a 50 percent chance to choose among both the constricting envelop system besides the spiral model to update the location of whales throughout optimization.

C. Phases of searching for prey and exploration :A strategy based on the diversity of the vector is employed in the hunt for prey. Furthermore, humpback whales seem to be randomly arranged when seen in relation to one another's positions.

$$\vec{H} = \left| y.\vec{C}_{rand} - \vec{C} \right| \tag{3.7}$$

and

$$C(t+1) = C_{rand} - y1.\vec{H}/Correlation \tag{3.8}$$

In the above equation, the arbitrary value is represented viz C_{rand}. The random number $rand$ and its selection are determined by the PSO algorithm's velocity value.

3.9.3 Velocity updating of the PSO algorithm

In the most common version of PSO, the velocity vector for a particle is maintained in accordance with the values of G_{best} and P_{best}, respectively. This is the methodology for revising the agent's (particle) velocity and location in the PSO: The agent's velocity and location are updated by the below Eqn. 3.9 correspondingly,

$$v_i(t+1) = \ v_i(t) + b_1 rand(Pbest(t) - r_i(t)) \ + \ b_2 rand(Gbest - r_i(t)) \tag{3.9}$$

$$r_i(t+1) = r_i(t) + v_i(t+1) \tag{3.10}$$

Where, the particle velocity referred by v_i and r_i designates the current particle, rand denotes an arbitrary or random number ranging from (0, 1), $b_1,\ b_2$ are learning factor, typically $b_1 = b_2 = 2$ This feature $|y\bar{1}| > 1$ highlights the research and makes it possible for MWO to locate the best global solution and update the searcher's location $|y\bar{1}| < 1$. Furthermore, to maximize efficiency in their quest for food, the whales move in short increments towards the prey in order to study the search zone effectively.

3.9.4 Criteria for terminating a contract

In light of all of this approach, it selects the ideal threshold with the best stability. Once the ideal results have been achieved, the procedure will be followed, but every position will be reviewed periodically. The search agent may then update its location in the vicinity of the current best value, which mimics surrounding the prey more effectively and allow for efficient updates. It therefore enables search agent respectively to update its location in the best value neighbourhood and mimics a more effective encircling of the prey.

3.10 Experimental Results Analysis: Privacy Preserved E-Health Care Data

The ideal threshold values in the privacy-preserving algorithm are determined with the help of the MWO algorithm, and then the health care data of patients is safeguarded from the acquired datasets with the help of the MWO method. The more precise safety of medical data in both datasets is attained (Heart disease and Wisconsin breast cancer). Finally, the suggested K-medoid algorithm combined with the MWO method achieves the highest possible accuracy. The efficiency of the K-medoid clustering method described in the results analysis section is discussed in further detail. The k- Medoid method is used to analyse the efficiency of the system based on the MWO. In order to protect the information of patients, the suggested solution employs two databases. This simulation method is carried out using Java Development Kit 1.4, which has RAM as 4GB and an Intel i5 CPU. The analysis of the precision and similarity measures is shown in the subsequent section. The hybrid optimization technique proved its efficacy in getting more accuracy compared to whale optimization algorithm.

3.10.1 The Dataset Description

Two medical datasets are used in the proposed clustering and security process: In the UCI repository of machine learning, there is the dataset of Wisconsin breast cancer and also the dataset of heart disease. The K-medoid with MWO technique is used to aggregate and protect the medical itemsets derived from these two datasets.

A. Breast Cancer Dataset: A major aspect of the Wisconsin breast cancer dataset is its use of digitized Fine Needle Aspirate (FNA) readings to establish significance. The image displays the various cell nuclei characteristics in the dataset. The description

Table 3.4 The Wisconsin vs Heart Disease Dataset

Dataset	Threshold	Time	Memory
WBCD	0.25	541236	4144544
	0.5	423542	3654874
	0.75	231458	2312454
HDD	0.25	18523	3612457
	0.5	22478	2814597
	0.75	26987	2145863

of this dataset has 10 characteristics, both nominal and linear, and 201 occurrences of one class, which is characterized by a total of 10 characteristics that are both linear and nominal in nature.

B. Heart Disease Dataset: This collection provides heart disease datasets that have 76 characteristics and 14 subsets. The values range from 0 to 4, which makes them integer values. Those investigations conducted with this database were able to discriminate the illness by the values 0 (missing) and 1 (appearance).

Table 3.5 Proposed MWO Algorithm: Preserved Itemsets

Dataset	Threshold	Number of Items	Number of Items Preserved
WBCD	0.25	997796	62012
	0.5	997796	84126
	0.75	997796	95562
HDD	0.25	1060	654
	0.5	1060	896
	0.75	1060	998

**Wisconsin breast cancer dataset: WBCD, *Heart Disease Dataset: HDD*

Table 2 describes the conserved itemsets from two datasets. The Wisconsin breast cancer dataset has 2 itemsets that are optimal using the suggested threshold optimization approach, while the heart disease dataset has 4 itemsets that are optimal (MWO algorithm). Increases in the threshold value result in an increase in the number of saved itemsets as well. However, at larger threshold values, memory allocation and execution time are minimal. For instance, in the Wisconsin breast cancer dataset, there are around 99,796,000 items. Using MWO, we can find 955,262 itemsets that have 23,158,544

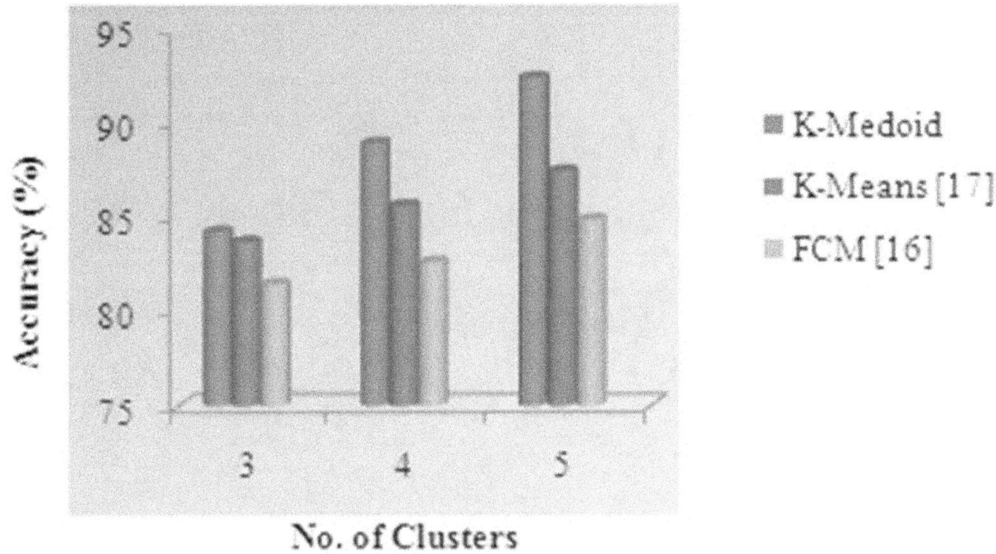

Fig. 3.4 Wisconsin Breast Cancer Dataset (WBCD)

items and 23,124,454 bytes of memory, using just 0.75 memory value settings.

3.10.2 Accuracy Analysis of Clustering

In the Fig.3.4 and Fig.3.5, it shows the results of the clustering accuracy of the dataset of breast cancer dataset and the dataset of heart disease were both used in this investigation.

Figures 3.4 and 3.5 show the results of the clustering accuracy study of the breast cancer dataset as well as the heart disease dataset, respectively. With the K-medoid method, the patient's data is grouped into distinct groups based on their characteristics. This Chapter examines and compares the presentation of the proposed K-medoid clustering method with the current K-means and FCM algorithms. For the breast cancer dataset, the greatest clustering accuracy is at cluster number 5, with a clustering accuracy of 92 percent. For the heart disease dataset, the maximum clustering accuracy is at cluster number 5, with a clustering accuracy of 94 percent.

3.10.3 Execution Time Analysis

In the Fig.3.6 and Fig.3.7, it is provided that a study of the execution time for threshold optimization with a variable number of clusters. The study is done on two datasets.

The first dataset is the Dataset of Wisconsin Breast Cancer, and the second dataset is

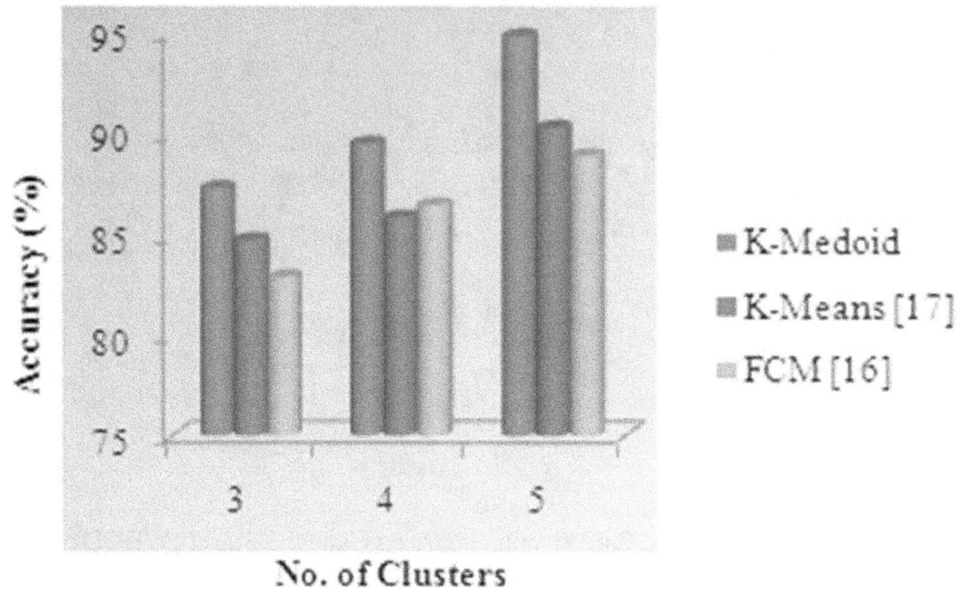

Fig. 3.5 The Heart Disease Dataset (HDD)

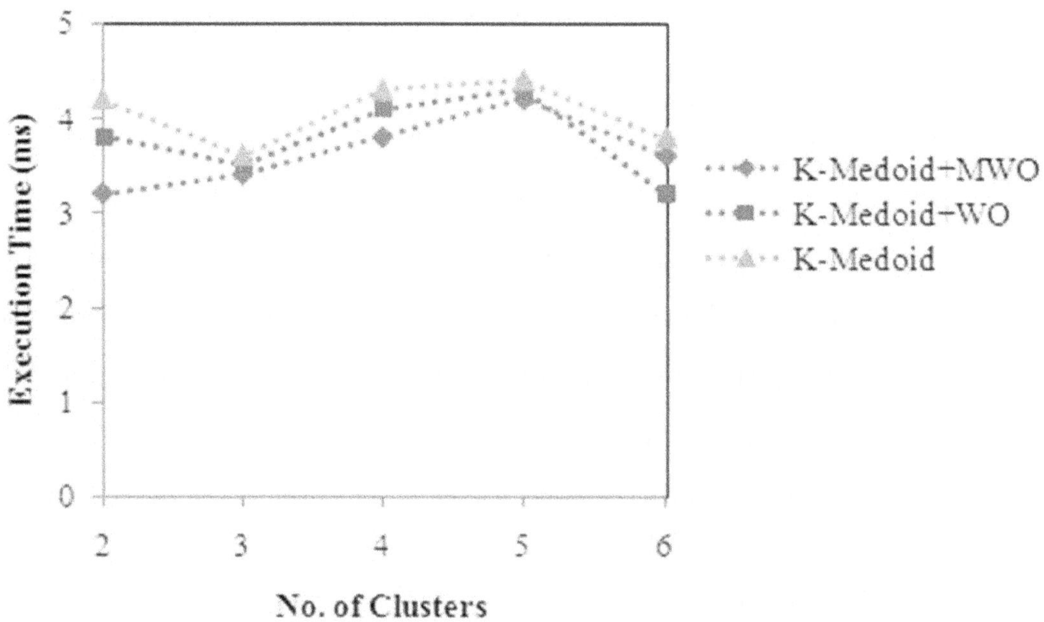

Fig. 3.6 The analysis of Wisconsin's breast cancer dataset shows how time influences certain outcomes

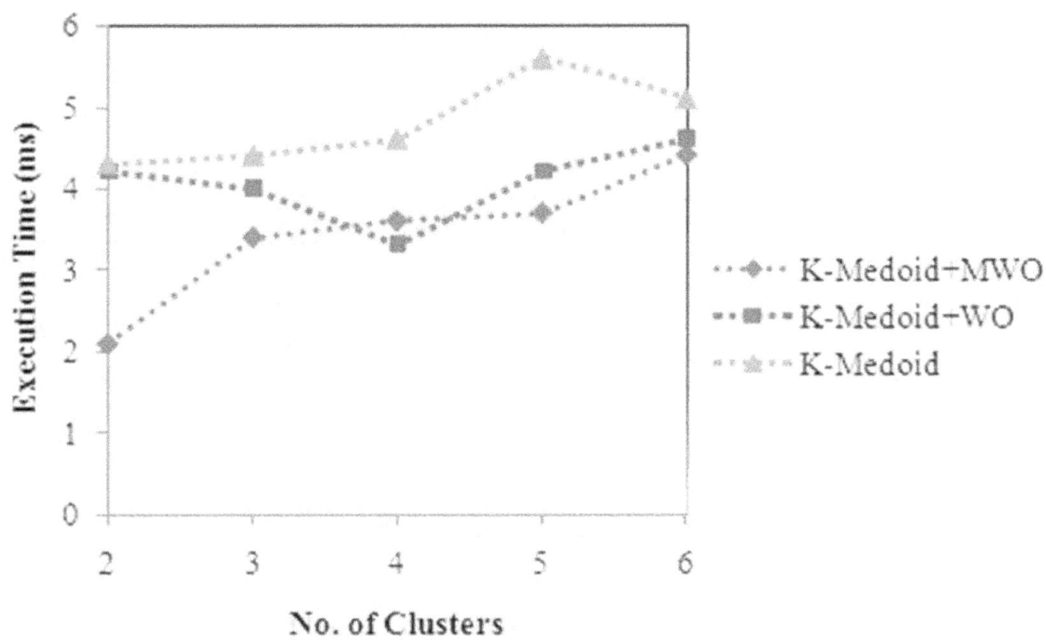

Fig. 3.7 Heart Disease Dataset Time Analysis

the Heart Disease Dataset. K-Medoid, K-Medoid with WO, and K-Medoid with MWO all had their computational processing time evaluated. Figure 3.6 shows that the K-Medoid with MWO requires 3.2 milliseconds to reach the optimum threshold of the cluster 2 and that the processing time for the other clusters is displayed on the dotted line graph as well. According to Figure 3.6, using K-medoid clustering and MWO, the optimum threshold is attained in 2.1 milliseconds. Thus, in comparison with other current techniques, the suggested model needs less time to optimize and cluster PMUIM thresholds.

3.10.4 Preserved Items

3.10.5 The Accuracy of Itemsets Privacy

The graph in Fig.3.10 illustrates how similar the preservation conditions are for the Breast Cancer Dataset of Wisconsin and the Dataset of heart disease. Fig.3.8 shows the number of items accessible for the saved medical articles. In this table, the total number of accessible item sets is 97796; the total number of items saved using the MWO method is 62012 at the 0.25 threshold. As can be seen in the image, the item sets

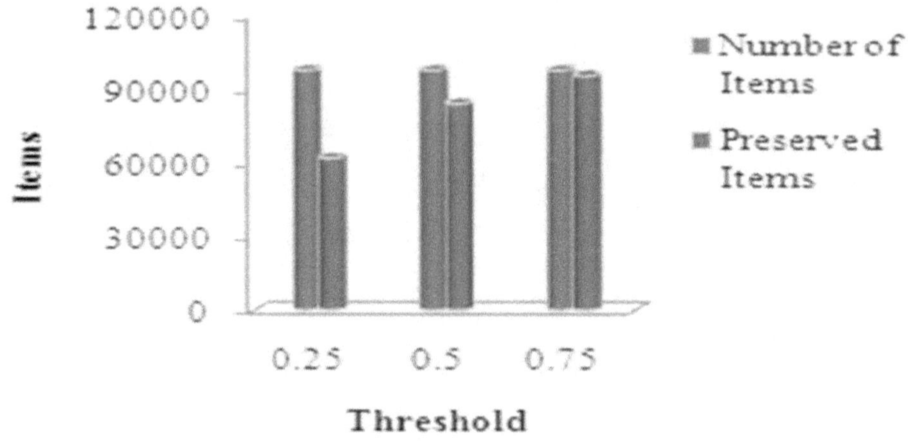

Fig. 3.8 Threshold Values of Preserved Items - WBCD

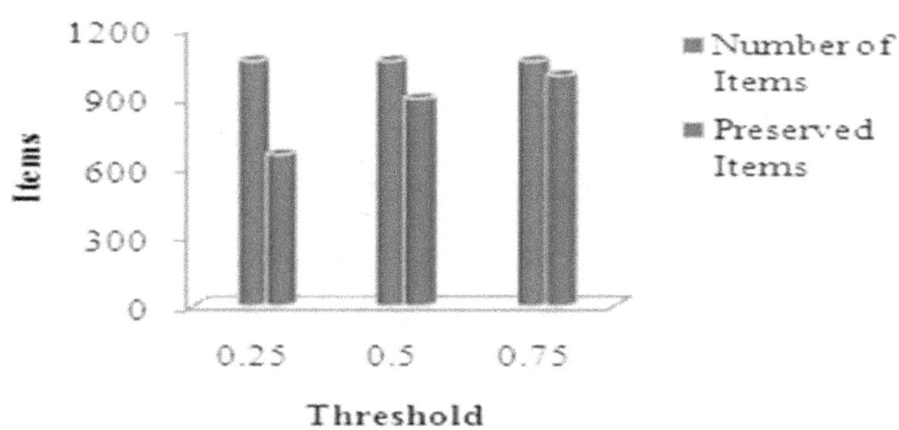

Fig. 3.9 Threshold Values of Preserved Items - HDD

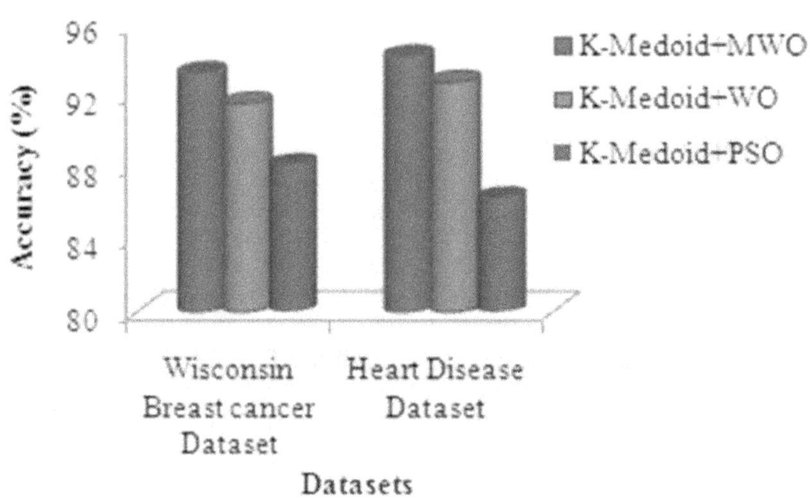

Fig. 3.10 Accuracy Analysis of Different Algorithms

for threshold values of 0.5 and 0.75 are shown on a graph. The number of things saved is likewise raised as the threshold value rises. The number of elements on the MWO algorithm is 1060 and 998, with 0.75 criteria as shown in Fig.3.9.

CHAPTER 4

PRIVACY PRESERVATION SCHEME FOR HEALTHCARE SYSTEMS USING MOWC AND OHUIS

4.1 Introduction

Electronic Health Records (EHR) is an electronic variety of patient health information (i.e. medical history) in an electronic setup. EHR is gradually becoming a reality in several developing nations. These documents can be pooled through a significantly associated network across various health solutions settings. EHR covers section and individual dimensions like age, weight, and charging information equally as essential signs, existing and previous medical history, household origins, medication, and level of sensitivities history, inoculation standing, study center examination results, radiology photos. EHR frameworks are intended in such a way which reduces paper medical documents to save information precisely and intelligibly in a computerized team. Health documents diminish the threat of information duplication as the files can be shared over the unique human solutions structures and also can be altered and updated. EHR programs are legally advantageous for the physicians, clients, and the facility the board authority. EHR framework can be useful in population-based investigations and also effective while separating clinical details for the assessment of prospective patterns and also lengthy term adjustments in a person.

Patients trust therapeutic administrations providers to make sure about the privacy and security of those disclosures. To keep up this confidence, express methods and establishment make sure open-minded protection is ensured. Patients, human administrations professionals, and evidence stockpiles are the standard segments of Electronic

Health Services (Fernández-Alemán *et al.*, 2013). Protection to long-term administration of patients' valuable records is provided given that information is looked after in the cloud, and it must be accepted by patients (Gøeg *et al.*, 2015). Furthermore, The EHR represents the strategy of remedial thought and make out of numerous sections, which includes data both in figured out data and regular language records such as medical technique programs, tests delineation, track history, discernments, clinical assessment, action by action status records in a crisis system diverse records (Palarivattom and Kochunni, 2015). It is trusted to find some significant and correct unquestionable conditions and standards that can visualize particular conditions or outcomes that can be fixed as procedures and recommendations for tolerant success of data making use of the record sets and fields (attributes) of the database (Pirbhulal *et al.*, 2019). Section 4.3 describes related background of the work and Section 4.4 discusses significance and scope of the work, Section 4.5 describes security and privacy prerequisites,Section 4.6 presents proposed model, Section 4.7 presents experimental results, Section 4.8 presents conclusion and Section 4.9 projects future scope of the work.

4.2 **Purposes**

4.2.1 Global e-health setup and execution

1. In Canada drove activity in 2001 to improve the framework of its medical services. In 2015, EHRs were generated 91 percent of Canadians; moreover, around 91,000 clinicians are situated consuming the EHR framework effectively.
2. In the year 2001, England inaugurated a national task termed the "National Plan for IT" (NPfIT) for revolutionizing the structure of its medical services. Further down the task, 96 percent of the people protected by the EHR structure.
3. In Germany, regarding 90 percent of medical professionals in the private method are expending EHR structures. Peaceful protection has been certain sufficient consideration in the task.
4. New Zealand has completed an EHR selection rate of 97 percent. Currently,

there is no focal or single EHR framework. However, a shared EHR framework, which can be received from any kind of compound the nation over. The nation is meaning to have a singular EHR framework by mid-2018.

5. In the United States, 76 percent of typical emergency centres and 97 percent of the intense consideration clinical facilities have been ensured EHR frameworks by 2014.

4.2.2 National e-health setup and execution

1. India has a combined medicinal services structure, including a huge variety of centre's run by both the State Government and supervised by Central Government equally as the exclusive part.
2. Enhanced use of technological innovation in health scenarios is one of the vital mechanisms of electronically accessed "digital India" tasks by the Government of India.
3. In the year 2013, Government of India has begun the task of the usage of information development in the human services structure, and for that purpose, Ministry of Health and also Family Welfare came up with EHR standards and criteria for India.
4. There are significant private associations like Fortis Apollo, Max, Sir Ganga Ram, as well as Sankara Nethralaya that have gotten digital documents while the remainder is stagnant in the organizing stage.
5. The Government of India has increased its strategy to \$ 19.2 billion to boost the government-mandated insurance framework by utilizing information technology ideas.

4.2.3 Worldwide standard EHR design and Implementations

The European Union's goal is to impart patients' EHR information to the theme of "Free Movement" and also of attaining top worth and well-organized human solutions

(Banaee *et al.*, 2013). In any case, there has remained alongside no motion in policy development, including the various essential safety and security issues elevated by action from a great extent is disengaged, paper-based health record structure is incorporated and digitally accessed (Fernández-Alemán *et al.*, 2013). Additionally, the breakthroughs in Information and also Communications platforms partake prompted a scenario wherein patients' health-related details are dealing with innovative safety as well as protection implications (Gøeg *et al.*, 2015). There are three crucial safety purposes (Goksen *et al.*, 2011): uprightness, accessibility, and secrecy of CIA.

The security and safety of individual data are basic in the health field, just as it is in this way expected to make certain the CIA of individual health info. As per the ISO EN13606 ordinary (Meri *et al.*, 2019), secrecy portrays the "strategy that verifies that data is effectively open just to those approved to approach it". Security alludes to the obligation to guarantee that evidence is precise and isn't changed in an unapproved style. The trustworthiness of safety and health data must, in this manner, be ensured to make certain individual security, and one significant piece of this resistance is that of certifying the data whole life cycle is clear and auditable. The calendar alludes to the "working of having the option to be utilized and realistic upon request by a licensed substance". The openness of health subtleties is, in like manner, vital to dependable medicinal services dissemination. The security and health informatics frameworks need to stay utilitarian when confronted with normal cataclysms, framework disappointments, and forswearing of administration ambushes. Health and security additionally involve responsibility, which indicates people's right to ask for what reason something has occurred.

Health data is additionally viewed by numerous individuals as being among one of the most important secrets of a wide range of individual subtleties. Protecting this security is thus significant if the individual security of subjects of care is to be kept. Individual security incorporates availability control against any occasion not approved to get to the data and consumes determined as "the case of foundations, people, or groups to distinguish on their own when, how, and what exactly level insights about them is associated to other people" (Metsker *et al.*, 2017). Protection in EHRs may be compromised by spyware or worms, software hackers, and software viruses. Various reports of unintended misfortune or the thievery of touchy logical data have shown up in the modern world (Mills, 2019), (Miotto *et al.*, 2016). Understanding the safety and protection qualities that EHR frameworks have might be basic if these dangers remain to be looked just as processes to build the data resistance of EHRs are to be received. As per contemplates acted in a few countries, issues relating to information health and security and individual protection have appeared. An ongoing examination study evaluated that every year there exists 25 million obliged approvals for the revelation of health and health records in the USA (Ogbuabor and Ugwoke, 2018). In research contemplates acted in Denmark, Switzerland, Germany just as New Zealand (Palarivattom and Kochunni, 2015, Pirbhulal *et al.*, 2019, Sahney and Sharma, 2018), members expressed user data assurance worries as apprehensions of EHR. Residents are furthermore conscious of the impending dangers that mutual EHRs may partake.

In Austria, people can likewise choose if their health-connected information ought to be imparted to foundations and medicinal services professionals. In request to relieve these worries, associations, such as the Certification Commission for Healthcare Information Technology (CCHIT), utilize an affirmed program that conceals a difficult assessment of, in addition to other belongings, safe keeping aspects dependent on existing experiments (Shen *et al.*, 2019), which is generally appropriate for the USA. CCHIT has remained authorizing EHR cutting edge innovation, given that 2006. Giving contact to EHRs is a fundamental next activity in turning on customers in their consideration and educating the health and health-related framework (Sonnati, 2017). This opens spic and span security and security endangers. There is a real issue seeing the two individuals'just as elements' benefit access to degrees to people's EHRs.

A person's EHR could be effectively open just as divided from various sites (by looking at different doctors' workplaces, clinical offices, transporters, etc). However, safety and security imperfections in a few of these frameworks could divulge the data of unsanctioned people or organizations, and health information requires protection from controls, unapproved access, and abuses, which incorporates considering protection, validity, check, accessibility, and duty issues (Vimalachandran *et al.*, 2016, Wang *et al.*, 2019). EHRs also experience complications in saving data of individual and privacy-related issues (Yüksel *et al.*, 2017), to the degree that the administration workforce may access data without the patient's particular approval (Zhu *et al.*, 2019).

4.3 Background

This area offers a background of some number of articles of journals, seminar proceedings, documents depending on various safety and security techniques, as well as elements being made use of in e-Health. In 2016, (Altamimi, 2016) as well as Altamimi checked out the safety and security as well as personal issues of privacy in e-healthcare equipment's. To get improvement in nature of health care administration, many EHRs could be certainly shared by an array of customers. This leads to huge personal privacy problems are routed to make use of EHR practical. In fact, the continuous expedition in norms of planning and standards make concerning security as well as personal privacy in EHR frameworks compulsory, it has still not entirely cleared out the personal privacy difficulties.

Healthcare deploying of the substantial information and examination were investigated by (Sonnati, 2017). Huge data triggers us to take a look at as well as re-create many regions not constricted to direction, health, and wellness, as well as legislation. The emphasis is to stress using the big information that is being done away with at all times reflecting in the history. Nevertheless, this is an optimal possibility to highlight on the examination to enhance the prescription and management.

An assessment concerning demonstrating information stream inside restorative administrations was illustrated in (Boddy *et al.*, 2019). Privileged the therapeutic data space, there is a prolonged requirement for noteworthy and also precise data. People ought to be sure that their profile information is being put away safely and securely. Especially, a portrayal of Transmission Control Protocol (TCP) connection affiliations is progressed, as an evaluation worrying the information multiangled nature and also client document interactions inside health care structures.

(Vimalachandran *et al.*, 2016) recommended an Ensuring Data Stability in Electronic Health And Wellness Records in 2016. Data credibility is the substantial pushed in EHR and is a basic item of organization movement whatsoever dimensions. Survey and deal with the effect of information uprightness of the usage of the EHR as well as its associated issues. An ideal method had been used to protect the decency in EHR structures. To dismember and also evaluate information decency, one of the dominant scientific systems was taken into consideration for the investigation.

(Gøeg *et al.*, 2015) checked out the Clustering scientific designs from close by electronic records in 2015. Scientific clinical designs in digital wellness records are usually passed on as layouts that help the various professional work kinds in which the system is utilized. The styles are much of the time organized, making use of the close by as against to conventional information designs and wording, which ruins inter-operability of semantic nature. In this way, it can provide a critical gadget to harmonization equally as the company of scientific models.

(Cebul *et al.*, 2011) had done a good investigation on topic of health electronic records and quality of diabetic treatment in 2011. We glanced at the establishment of a renovation in benchmarks of quality for diabetes, making use of EHRs with those at manage making use of paper documents. All having a go at, consisting of numerous protection web fundamental idea practices, had a chance with a local top-worth group and also straight proclaimed execution. (Miotto *et al.*, 2016) talked about the anticipation of the patients' future from health electronic records in 2016. Our results on a really basic level beat those grown, making use of depictions reliant on raw EHR information as well as elective component finding out techniques. The need for implementation for genuine diabetes mellitus, schizophrenia, and various hatreds was among the top list. These happenings reveal that profound discoveries associated with EHRs can determine figured out representations that provide boosted medical opinions and give an AI structure to expanding scientific determination systems.

4.4 Significance, Scope, and Definitions

Information technology has actually prompted a particularly major enhancement in the health care region called electronic health care (e-Health). No matter, with perceivable as well as different beneficial scenarios inevitable from electronic health in handling of distributed nature, its complete utilization is so far being obstructed by obstacles of protection equally as personal privacy issues. In the projected work, more than one dataset is taken into consideration for the private investigation, and also it has actually been taken from the University of California at Irvine (UCI) AI vault. The dataset comprises a lot of information in terms of rows in lakhs; since the professional datasets, the particular sicknesses are set up or bunched by their degree of the condition. For clustering, an imaginative model has offered, for example, Multi-Objective Weighted Clustering (MOWC). Additionally, to upgrade the privacy-safeguarding for grouped information (individual's document), it is made certain around the cloud with the assistance of item set mining which functions depending on the cut-off value. The perfect restriction value is picked by utilizing compromise streamlining; this computation emulates the conduct of two distinct multitudes, for instance, Salp Swarm and Particle Swarm Optimization (SS-PSO). Apparently, the end results demonstrated that the mix of MOWC as well as item set mining estimations boost the precision of e-health services information privacy in entire datasets.

4.5 Security and Privacy Prerequisites for e-Healthcare model

Electric Health Records (EHR) is a state-of-the-art amount assortment of patient health and health information in an electric configuration. EHR ends up being increasingly being actually performed in a lot of developing nations (Abul-Husn and Kenny, 2019), these EHRs are used in a lot of developed countries in nursing establishments, health facilities, as well as other associated health organization bodies. Moreover, the purpose of insurance policy coverage administrations is regularly to render unbiased, effective, and also tough services that are individual. EHRs offer a focal job in interfacing digital point to consider institutions, which have been realized as parties of private insurance policy coverage professional (HCPs) just that provides an alarm to close customers.

Patients trust therapeutic administrations providers to make sure about the defence of those disclosures. To keep up this confidence, express methods and establishment make sure open-minded protection is ensured (Banaee *et al.*, 2013). Patients, human administrations professionals, and evidence stockpiles are the standard segments of Electronic Health Services (Fernández-Alemán *et al.*, 2013). Protected long-term administration of patient valuable records given that information is looked after in the cloud, and it must be accepted on by patients (Ogbuabor and Ugwoke, 2018). Apparently, the EHR represents the strategy of remedial thought and make out of numerous sections which included data both in figured out data and regular records of language, for example, medical technique programs, tests delineation, track history, discernment's, clinical assessment, action by action status records in a crisis system diverse records (Palarivattom and Kochunni, 2015). It is trusted to find some significant and correct unquestionable conditions and standards that can visualize particular conditions, conditions, or outcomes that can be fixed as procedures and recommendations for tolerant success data making use of the record sets and fields (attributes) of the database (Pirbhulal *et al.*, 2019).

E-health phase increases protection as well as privacy issues as a result of the sensitive idea of wellness information (Goksen *et al.*, 2011). Crucial concerns of protection as well as privacy in the e-Health cloud integrate enigma, reliability, approval, obtain the possibility to control, as well as non-renouncement (Meri *et al.*, 2019). Safety and also personal privacy in EHRs can be really threatened by software application designers, contamination's, as well as worms. Different reports of unintended bad luck or the burglary of delicate professional information have shown up of late (Metsker *et al.*, 2017). There is a big examination in substantial data for information confirmation, putting together, and the sharing of wellness details and also data use (Mills, 2019). A couple of scientists use dispersed signing up as well as conveyed stockpiling to deal with this issue. Regardless, this brings an additional deterrent to the usage of the e-Health system

(Miotto *et al.*, 2016) Workouts Research methodologies are a huge and worthwhile gadget for taking account of a wide extent of health care problems (Ogbuabor and Ugwoke, 2018).

The e-health stage increases safety, as well as assurance issues due to the delicate concept of wellness info (Goksen *et al.*, 2011). Fundamental concerns of protection as well as a guarantee in the e-Health cloud join unidentified, dependability, recommendation; discover the opportunity to regulate, and non-denial (Meri *et al.*, 2019). Safety and security and also guarantee in EHRs can be overturned by setting developers, infection, and also Spywares, and also worms. Various records of an unintended problem or the thievery of touchy professional info have shown up lately (Metsker *et al.*, 2017). There is a massive test in enormous details to the level of information affirmation, accumulating, and sharing of health and wellness data as well as information utilize (Mills, 2019). Numerous specialists use appropriate figuring and circled around storage to handle this problem. No matter, this conveys another impediment to the utilization of the e-Health framework (Miotto *et al.*, 2016). Activities research study techniques are a remarkable and also productive means for handling a vast level of professional addition issues (Ogbuabor and Ugwoke, 2018). Prolonged availability of improved interdisciplinary document, patient information, upgraded intelligence of idea, decipherability, improved paperwork, decreased duplication, and boosted rate are numerous inclinations (Boddy *et al.*, 2019). The openness of huge proportions of modernized data structured in EHR, got together with the renovation of rising information in health systems, is launching new exam roads and also open entrance ways for boosting patient's health and wellness document privacy (Cebul *et al.*, 2011). In addition, the nonstop use of the Internet of Medical Points without recalling the patient's data order as well as safety will certainly encourage unauthorized people to listen to discreetly restorative data during its transmission between far-off devices (Sahney and Sharma, 2018). Multi-Objective Weighted Clustering approaches have actually been comprehensively made use of in the recent years for displaying health care structures and also exploring their introduction. To this factor, beginning late such Multi-Objective Weighted Clustering strategies have been

Table 4.1 Privacy Preservation –A Comparative analysis

Investigation Manuscript	Proposed model	Patient Privacy Level	Associating PHR for clinical research	Privatized data searching
(Deng et al., 2020).	Partition of vertical data and hybrid privatized Data searching	Robust on data identity, general data	No	Yes
(Riedl et al., 2008).	Assigning new PSN to a PHI for every session. preventing correlation of two PSNs originating from single PHI	Fragile on identification	No	No
(Xu et al., 2014).	Revolutionary the idea of data preservation with data partitioning; multiple stalls with inadequate patient info'	Fragile on data	No	No
Riedl et al., 2008).	Assigning first-hand PSN to a PHI for every single gathering avoiding association of two PSNs instigating from solitary PHI	Fragile on identification	No	No
(Aamot et al., 2013).	dentifies glitches connected with Pseudo preservation method familiarized by (Riedl et al., 2008)	Fragile on identification	No	No
(Agrawal et al., 2003)	Hippocratic data base for permissible and virtuous e- Healthcare acquiescence	Robust on data identity, general data	No	Yes
(Alhaqbani and Fidge, 2008)	EHR connecting to distinct EMRs	Fragile on identification	Yes	No

accompanied enhancement frameworks (Mirjalili *et al.*, 2017).

4.6 The Proposed Mining Model of Data: Multi-Objective Weighted Clustering

The MOWC algorithm aims for splitting the records of health care reliant on the patient's illnesses. MOWCA is to be contingent on three basic constraints, which are: degree variance, the relation, and distance among data or datasets and its neighbors and finally the distance of mission. By employing the MOWC algorithm, the harshness of diseases with the least combined weight is elected as a head of group. The constraints used in algorithm of MOWC are contemplated in Table 4.2.

Parameters	Description
D	Data or Data Sets
$D_{diff\,i}$	Degree Difference of data i that is the difference between the degree of data i (Number of datasets within its communication radius R) and a predefined ideal node number Max in a cluster.
D_{ci}	Sum of distances between data i and its Neighbors.
D_{mi}	Mission distance of healthcare data i.
$C_{v(i)}$	$C_{v(i)} = \left\{ j \in C \middle/ \left[\left(X_i^{op} - X_j^{op}\right)^2 + \left(Y_i^{op} - Y_j^{op}\right)^2 \right]^{1/2} \leq R \right\}$ the neighbour set of each data i
M_i, N_i	The current position of data i
M_i^{op}, N_i^{op}	The optimal position of data i
M_i^c, N_i^c	Communication position of data i
d_i	degree of each data i

Fig. 4.1 The parameters used in the MOWC algorithm

4.6.1 Organization: E-Healthcare Architecture

The projected e-Healthcare architecture focuses on enhancing the preservation of privacy in data of e-health by doing of clustering of data and algorithm of mining of item sets. Here, two distinct datasets of health care are taken from the UCI repository of machine learning. The grouped information of health is protected by the creative technique

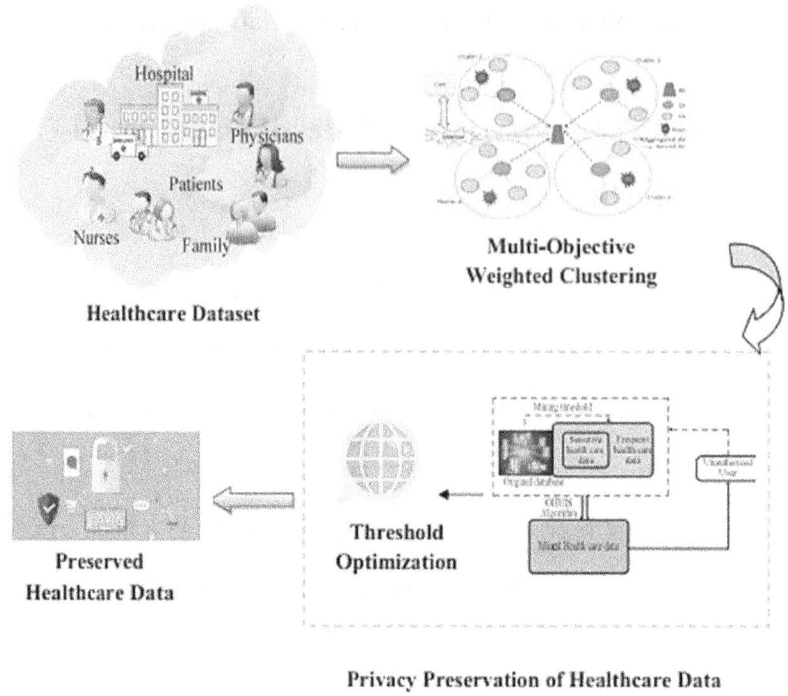

Fig. 4.2 Architecture: Privacy Preservation of E-healthcare Data

i.e. Optimal High Utility Item Selector (OHUIS), where the ideal value of threshold is computed by the meta-heuristic algorithm called SS_PSO. The architecture of the presented work is illustrated in Fig.4.1. The implementation for preservation of privacy of healthcare data consists of three phases.

4.6.2 Module Segment - 1: Data of healthcare

In the anticipated work, two data sets of health-care remain concerned from the UCI repository of machine learning to validate the electronic social data of insurance in the cloud. The databases of reality are huge and are aggregated from heterogeneous and multiple sources. Next, the clustering is crucial as manifested by the patient's disease by the proposed clustering technique.

4.6.3 Module Segment - 2: Multi-Objective Weighted Clustering

The e-Health care details have thousands or lakhs of patients' info for various ailments. This is taken into consideration, and databases are put together right into the kind of little various subgroups, which are called as clustered teams. The mean towards clustering is to set the sickness class into a lot of information focuses (healthcare info of each individual's) based on the resemblance measure. Every detail factor inside a similar lot has more prominent resemblances when contrasted with the info straight having a place towards one more team. Henceforth, clustering is applied in the privacy safeguarding of datasets where the information is constructed dependent on the infection.

We presume that a regular set of n data or datasets is organised in a defined area of concentration to administer a given corporeal occurrence. The pattern setup is characterised by the graph G = (C, E), wherein C= 1, 2, . . . , n is a set of n data or datasets, and E ⊂ C ×C is the set of the relation among the various data or datasets. The $C_{v(i)}$is the neighbour set of the data i. Let overall time required to reach the distance from data x to data y is represented as $f_{xy}(D_{ci})$ an optimal or mission time is required for the data set x is represented as $f_i(D_{mi})$, then the optimal locations of data or data sets by solving the optimization problem described as trails,

$$\min f(x,y) = \sum_{i} f_i(D_{mi}) + \sum_{i} \sum_{y \in C_{v(i)}} C_{v(i)} \cdot f_{xy}(D_{ci}) \tag{4.1}$$

Subject of,

$$(M_i,\ N_i) \in R,\ \forall (i,i) \in D \times C_{v(i)} \tag{4.2}$$

We put the solved equation as,

$$f_i(D_{mi}) = \sum_{y \in C_{v(i)}} f_{xy}(D_{ci}) \tag{4.3}$$

$$D = (M_1,\ N_1, M_2,\ N_2, ...M_i,\ N_i) \tag{4.4}$$

$$F_1(M_i,\, N_i) = \sum_i f_{xy}(D_{ci}) \tag{4.5}$$

$$F_2(M_i,\, N_i) = \sum_i f_{xy}(D_{mi}) \tag{4.6}$$

$$F(D) = f(M_i,\, N_i) \tag{4.7}$$

Therefore Eqn. 4.1 becomes,

$$\min F(D) = F_1(D) + F_2(D) \tag{4.8}$$

4.6.3.1 Algorithm 4.1: Computation of optimal position of Data *i*

Data:

$$(M_i,\, N_i), (M_i{}^c,\, N_i{}^c), i \in D$$

Result:$(M_i{}^{op},\, N_i{}^{op}), i \in D$

1. Start

. a. Initialization of total dataset D_i ;

. b. EvaluateD_i using the function F ;

2. While No convergence

. a. do

. i. $(D_i)'$: = Selection of cluster head in D_i;

. ii. $(D_i)'$: = Integrate the crossing operator on $(D_i)'$;

. iii. $(D_i)'$: = Integrate the mutation operator on $(D_i)'$;

. iv. D_i : = Replace old cluster head of D_i by their descendants in $(D_i)'$;

3. Evaluate D_i using the function F;

4. End

The optimal position computation algorithm is given in algorithm 4.1 above.The proposed MOWC algorithm is initiated by generating the initial data sets D_i (Several values of Eqn. 4.3 as well as assessing the version of all given datasets and various values of M_i,N_i). Formerly the discrete datasets are arbitrarily nominated for reproduction, rendering to the norms of endurance of the fittest. Subsequently, the child datasets (or descendants) are produced relating the subsequent dual genetic operators: mutation and crossover. In fact, those child datasets are migrated to new datasets $(D_i)'$ and substituted in entire or in part by the child datasets of preceding generations. Furthermore, the new datasets take above from one generation to the next in anticipation of attaining the ending criterion. This can be performed up to several simulations, the value of Energy efficiency factor (e) = 0.0001 as ending criterion comparatively to the calculation step.

After computing the optimal location for each given data or datasets i using MOWC, in this segment represents the central phase of the specified model. The algorithm 4.2 describes multi-objective weighted clustering algorithm. Certainly, to elucidate the energy efficiency limitation and to enhance the resources in very complicated data or datasets, we established the MOWC algorithm depend on the subsequent constraints:

DM_i : Mission distance of data or datasets i.

DC_i : Sum of distances among data or datasets i and its neighbours.

D_{diff_i} : Degree Difference of data or datasets i that is the variance among the degree of data or datasets i .

EC_i : Energy Efficiency Characteristic of data or datasets i.

4.6.3.2 Algorithm 4.2: Multi-Objective Weighted Clustering

1) Initial computation of EC_i

$$EC_i = (R_i \times A)/E_i \tag{4.9}$$

R_i Rate of the relationship between the datasets, the required energy or computation power for initial dataset identification E_i. The constant factor A.

2) Determine the neighbor set $C_{v(i)}$ of each dataset i, $C_{v(i)}$ is described as,

$$C_{v(i)} = \left\{ j \in C / \left[(X_i^{op} - X_j^{op})^2 + (Y_i^{op} - Y_j^{op})^2 \right]^{1/2} \leq R \right\} \qquad (4.10)$$

Compute the degree d_i of each dataset i is defined as $d_i = \text{Card}(C_{v(i)})$

3) Compute the degree alteration of dataset i by implementing this below formulation,

$$D_i = |d_i - Max| \qquad (4.11)$$

4) By computing, expression of Eqn. 4.8 is described as: The degree difference, and the quantity of the distances among data i and its neighbors are computed by the Eqn. 4.11 and Eqn. 4.12.

$$D_{diff_i} = |d_i - Maximum| \qquad (4.12)$$

Where,

$d_i = Card(C_{v(i)}$

$$D_{ci} = \sum_{j \in C_{v(i)}} \left[(M_i^{op} - M_j^{op})^2 + (N_i^{op} - N_j^{op})^2 \right]^{1/2} \qquad (4.13)$$

5) Compute the parameter DM_i which defines the distance between the ideal position of profile and the profile's position of mission data.

$$D_{mi} = \left[(M_i^{op} - M_j^{m})^2 + (N_i^{op} - N_j^{m})^2 \right]^{1/2} \qquad (4.14)$$

6) Compute the collective weight W_{ci}using the following eqn.4.15.

$$W_{ci} = w_1(D_{diff_i}) + w_2(D_{ci}) + w_3(D_{mi}) \quad (4.15)$$

The terms $w_1,\ w_2,\ w_3$ are different weights, and the amount of $w_1 +\ w_2 + w_3 = 1$

7) Choose the dataset with the least collective weight $C_{v(i)}$ as a cluster head.

8) Eradicate the cluster head that is selected and its neighbours from the group of unique dataset nodes.

9) Accomplish Steps 1–8 for the enduring datasets until the respective solitary is allotted to a cluster.

Subsequently to the implementation of these equations, all the points are re-clustered nto groups with concerned CHs. Then remove the head of the cluster that is selected and its adjoining members from the group of original data. The process is repeated for the enduring data till each one is associated to a cluster.

4.6.4 Optimal High Utility Item Selector

The primary goal of the Optimal High Utility Item Selector (OHUIS) formula is to choose the most ideal items coming from the sensitive item sets possessing the same very high energy utility value. The higher utility value product collections are actually hidden by customizing values of the frequency of things contained in the vulnerable item sets based on the minimum utility limit α. This concealing method is actually repeated until the whole sensitive thing collections electrical worth to end up being lower than the threshold limit α. This market value is actually acquired due to the planned protocol and the actions associated with OHUIS presented beneath. The process of exploration healthcare information is depicted in figure 4.5. The algorithm 4.3 describes

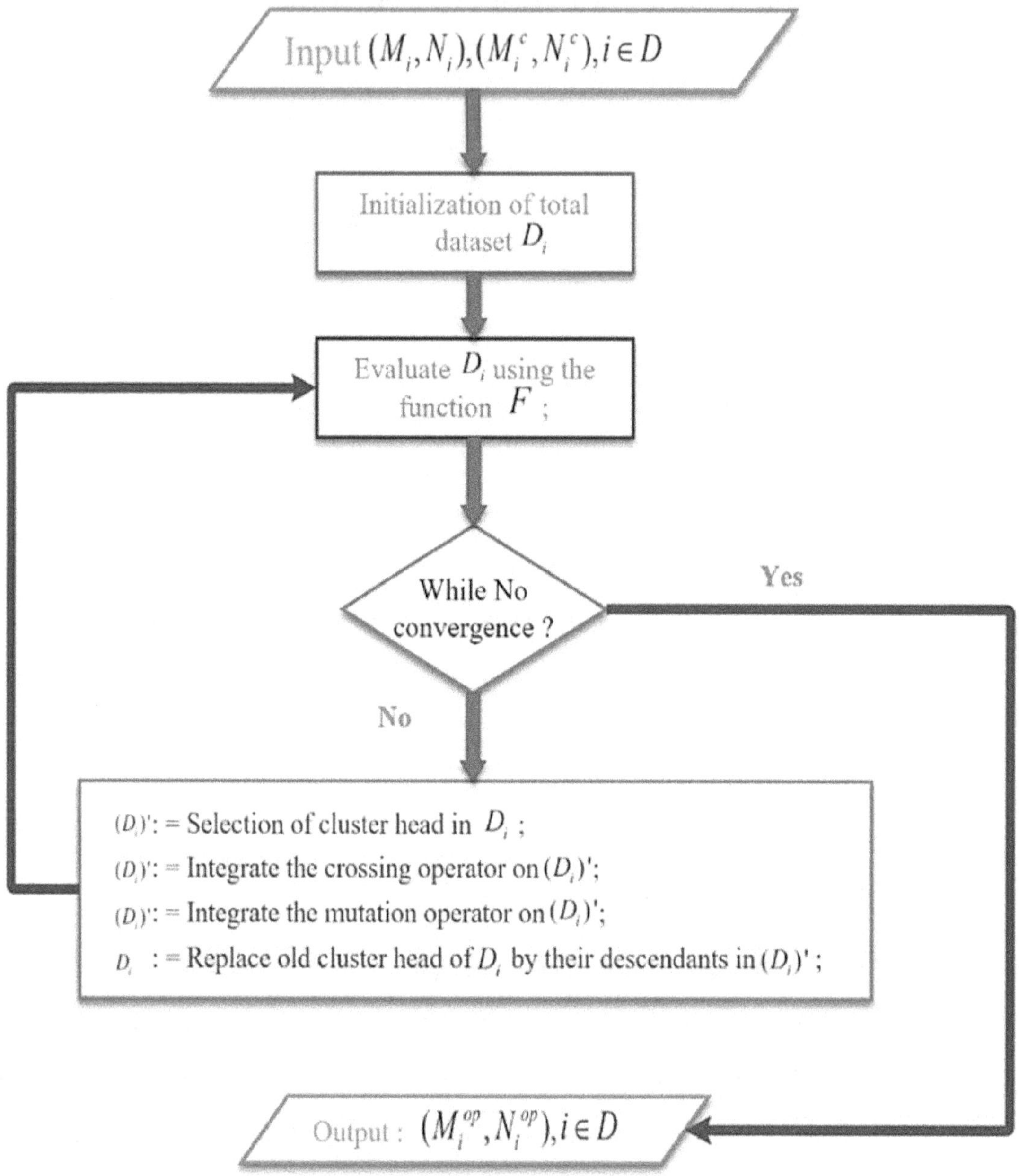

Fig. 4.3 Flowchart - Computation of optimal position of Data

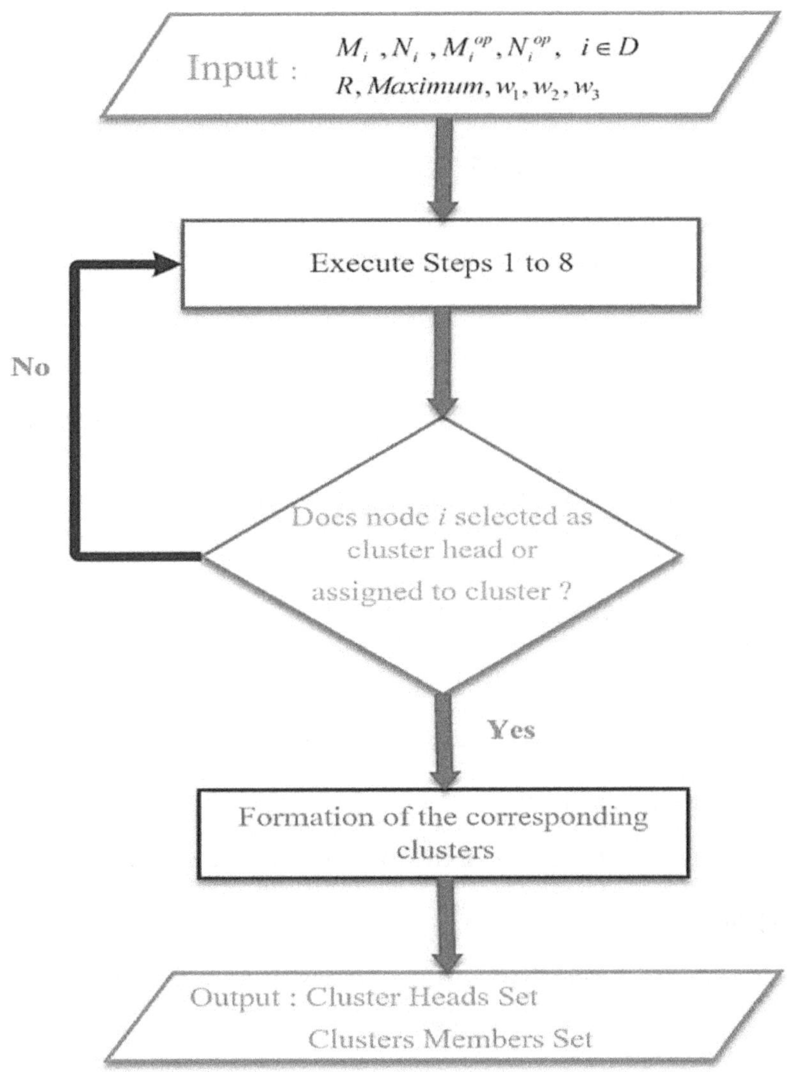

Fig. 4.4 Flow Chart - MOWC algorithm

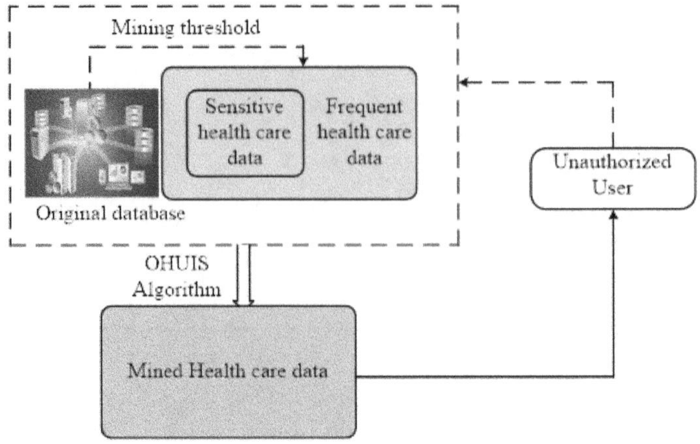

Fig. 4.5 Frequent Healthcare Data Mining

the novel optimal high utility item selector procedure.

4.6.4.1 Algorithm 4.3:Optimal High Utility Item Selector (OHUIS) (D, SI, T)

Input - D : Transactional Database, SI: Set of Sensitive Items, T: Initial Minimum Utility Threshold

Output- Sanitized Database and high utility item sets.

1. Start

2. Initialize D=Medical Database, T=Initial minimum utility threshold.

3. Repeat

4. Initialize the T value in SS_PSO to a value between 0 and 1.

5. Repeat

6. M = Project the database of transactions that are having high utility sensitive item sets.

7. Repeat

8. for a sensitive item set $S_g \in D$

9. $diff$ = utility $(S_g)^T$

10. if $S_g$contains two item sets $S_g \in (g_x, g_y)$ Then

11. if (utility $(g_x)^T$) = = (utility $(g_y)^T$) then

12. Select g_x, g_yitem frequency values $v(g_x)^T, v(g_y)^T$

13. Find the correct frequency value which optimizes accuracy using threshold T with SS_PSO model.

14. else

15. Select the high utility sensitive item set.

16. end if

17. end if

18. end for

19. Until all transactions in M are over.

20. Until every sensitive item set.

21. Until accuracy reaches high.

22. end OHUIS

4.6.5 Salp Swarm and Particle Swarm Optimization

Salp Swarm Optimization (SSO) is another optimization method (Mirjalili *et al.*, 2017) created to tackle various sorts of enhancement issues. Salps travel through contracting, so siphoning water through their jellied bodies change their positions. Salps, in seas, have swarm conduct called salp chain, this conduct may help salps in searching and improving development utilizing quick agreeable changes. It impersonates the conduct of Salps in nature; salps are a sort from the Salpidae's family, and they are barrel-formed tunicate of planktonic nature. Also, they are like jellyfishes in tissues, and moving conduct and their loads have a high water rate. In view of this lead, the author (Mirjalili *et al.*, 2017) displayed the salp chains in a numerical structure and tried it in optimization issues.

SSO starts by segmenting the populace into two events, in particular, the pioneer as well as the followers. The front salp of the chain is called the pioneer, as well as different salps are referred to as the supporters. The salps' setting is solved in n-dimensions which speak to the hunt area of an issue, and also n represents with the

Fig. 4.6 Frequent Healthcare Data Mining

challenging factors. These salps check for a food resource that shows the goal of the throng. The setting ought to be rejuvenated a lot of the time, so the coming with the condition is made use of to play out this task to the salp pioneer.

SSO starts by segmenting the populace into two events, in particular, the pioneer as well as the followers. The front salp of the chain is called the pioneer, as well as different salps are referred to as the supporters. The salps' setting is solved in n-dimensions which speak to the hunt area of an issue, and also n represents with the challenging factors. These salps check for a food resource that shows the goal of the throng. The setting ought to be rejuvenated a lot of the time, so the coming with the condition is made use of to play out this task to the salp pioneer.

$$W_j^1 = \{f_j + R_1 \left|(U_j - L_j)R_2 + L_j\right|\} \quad R_3 \geq 0 \tag{4.16}$$

$$W_j^1 = \{f_j - R_1 \left|(U_j - L_j)R_2 + L_j\right|\} \quad R_3 < 0 \tag{4.17}$$

where W_j^1 is the spot of the pioneer inside j^{th} dimensional measurement, where the food source in this dimensional measurement is f_j the upper and the lower limits $U_j \ and \ L_j$ individually. $R_1, R_2 \ and \ R_3$ are created arbitrarily in the range [0, 1] to keep up the hunt space. Likewise, the boundary is the significant coefficient in this calculation because of its job in the adjusting between the investigation stage and the misuse stage, and it is determined as follows:

$$R_1 = 2e^{-\left(\frac{4l}{L}\right)} \tag{4.18}$$

$$W_j^i = \frac{1}{2}\left(W_j^i + W_j^{i-1}\right) \tag{4.19}$$

Where l and L demonstrate the current cycle and the maximum iterations' number, separately, subsequent to refreshing the pioneer's position, the SSO begins to refresh the follower's position utilizing the above condition. The W_j^i is i^{th} the supporter position inside the jth dimensional measurement and i is more prominent than 1. The last strides of the SS are given in Algorithm 4.4.

4.6.5.1 Algorithm 4.4: Salp Swarm Optimization (SSO)

1. Prepare the population W
2. Replicate
3. Calculate the objective function for every solution Wi
4. Update the finest salp (solution) (f=Wb)
5. Update R1 using Eq.4.17
6. for i=1:N do
7. if i==1 then
8. Update the location of salp using Eq.4.16
9. else
10. Update the location of salp using Eq.4.17
11. end if
12. end for
13. until $(l < L)$
14. return the finest solution f.

Particle flock optimization (PSO) was introduced by Eberhart as well as Kennedy

(1995); it simulates the evolving of the expertise of social habits and imitates the team communication behaviour when sharing exclusive expertise concerned to searching, flocking, or migrating. This team and its participants represent a solution and hence are called a flock as well as particles, respectively. A bit counts on its experience and also neighbors to upgrade its setting. The swarm starts by developing a set of arbitrary particles in addition to producing their placements (W_i) and also velocity rate (v_i) in a dimensional measurementj^{th}. Afterwards, PSO begins its key loop to review each particle by executing a fitness feature; the outcome is gotten in touch with its ideal and worldwide ideal values. The equations (4.19) and (4.20) stand for the method that is used to upgrade the bits positions.

$$W_{ij}^{(l+1)} = W_{ij}^{(l)} + v_{ij}^{(l+1)} \tag{4.20}$$

$$v_{ij}^{(l+1)} = qv_{ij}^{(l)} + b_1 r_1 (W_{ij}^{p(l)} + W_{ij}^{(l)}) + b_2 r_2 (W_j^{g(l)} - W_{ij}^{(l)}) \tag{4.21}$$

where W_{ij} is the ith fragment position in the jth dimensional measurement, v_{ij} is the ith velocity in the jth dimensional measurement,l describes the present iteration,q specifies an inertia weight, and put on boost the rate of the population merging. The coefficients $b_1 \ and \ b_2$ are acceleration factors. $W_{ij}^{p(l)}$ describes the best preceding placement of bit i inj^{th} dimensional measurement; $W_j^{g(l)}$ specifies the universal best placement in j^{th} dimensional measurement. $r_1 \ and \ r_2$ are arbitrary parameters [0, 1]. The algorithm 4.5 describes the procedure for particle swarm optimization.

4.6.5.2 Algorithm 4.5:Particle Swarm Optimization(PSO)

1. Initialize the community of particles.
2. repeat
3. for i=1 to community Size, do

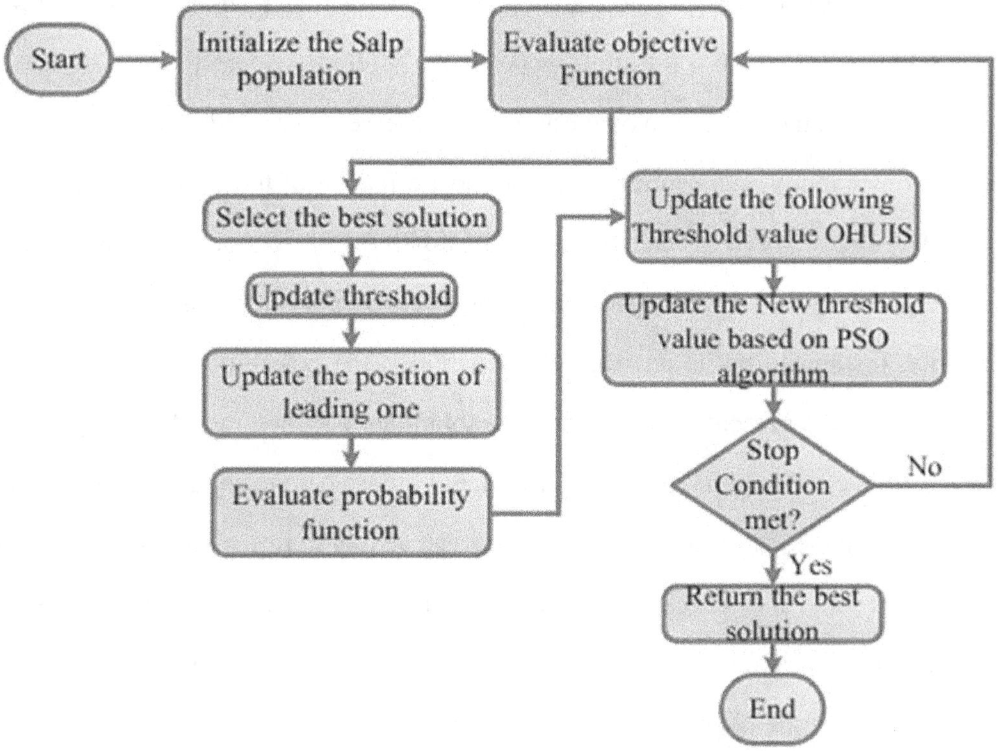

Fig. 4.7 Flowchart of SSPSO Algorithm

4. Compute the intended function value.
5. If the value of fitness is finer than the personal best one (PBest)
6. Then make the ongoing value as the new PBest.
7. Choose the particle of swarm with the finest fitness value of all the particles as GBest.
8. Compute a new velocity in coordination with Eq.(4.21).
9. Upgrade the particle's position in coordination with Eq. (4.20)
10. end for
11. Until termination criterion is met.

The fundamental framework of the SS calculation is changed by refining the rejuvenating period of the populace's setting. This change combines the upgraded tool of the PSO right into the fundamental assembly of the SS. This mix adds greater tractability to the SS in investigating the populace as well as assures its diverse variety, equally as reaches the optimum worth swiftly. When all is stated and done, the basic framework of the recommended SSPSO calculation is given in Fig. 4.7. Where the initial phase in the recommended SSPSO is to characterize the borders as well as produce the population, which speaks with a lot of answers for the offered concern (consist of selection). At that point, the discussion of every setup is assessed by refining the physical fitness work for everybody and also regulates the most effective of them. The succeeding phase in the suggested SSPSO calculation is to freshen the present populace by using one or the other the SS or PSO computation, which relies upon the nature of the physical fitness job (estimated by its possibility). Wherein if the possibility of health and suitability job, for the present setup, is extra notable than 0.5, then the SS, in any case, the PSO is utilized. From that point, the health and fitness help every arrangement is signed up, and also the very best setup is resolved succeeding to refreshing the people. The subsequent stage is to check in the occasion that the quit circumstances are fulfilled, at that point reappearance by the finest arrangement, all the same, rehash the past strides from registering the likelihood as far as feasible.

The SSPSO algorithm initiated by specifying the preliminary values of the SS as well as the PSO, after that SS creates an arbitrary population X of proportion N in dimensional measurement D, then SS computes the food health and fitness for every remedyW_i, i = 1, 2,., N. Nevertheless, before calculating the unbiased feature of objective function, each remedy W_i is transformed to a binary vector (which comprises only 1's 0's) conferring to the value of an arbitrary threshold limit ε [0, 1] utilizing the subsequent equation:

$$W_i(l+1) = \left\{ 1 \ if \frac{1}{1+e^{-W_i(l)}} > \varepsilon \ 0 \ otherwise \right\} \tag{4.22}$$

Only the W_i components that are equivalent to 1's are selected to stand for the designated features. Similarly, the supplementary elements are neglected, considering that they symbolize the unnecessary attributes.

4.7 Results and Discussion

The analysis of result area reviews the efficiency of the recommended MOWC clustering formula. The efficiency is analyzed and constructed upon the OHUIS formula together with limit optimization. The recommended strategy uses two datasets for the process of safeguarding patient's facts. This virtual procedure is carried out by JDK 11.0.7 with 16GB RAM as well as an i7 processor. The precision and also resemblance measure analysis are gone over in the section listed below. In the proposed healthcare information clustering and privacy and security procedure, a pair of datasets are actually gathered coming from the UCI library of machine learning that is actually: breast cancer data set of Wisconsin as well as dataset of heart disease. (a) The Wisconsin breast cancer dataset consists of 250 instances of one training class as well as 95 occasions of an additional class, and also it is actually portrayed through 10 qualities that are straight equally as plausible. (b) Secondly, the Heart problem dataset has 76 physiognomies as well as 14 subcategories. The high qualities are the entire amount based on 0 to 4. The hybrid optimization technique with novel clustering algorithm proved its efficacy in getting high accuracy compared to K-medoid and PMUIM algorithms.

Table 4.2 Distances of Euclidean from each point to updated centroids of cluster

Threshold	Dataset	Number of Items	Number of Items Preserved	Time(ms)	Memory (Bytes)
0.25	**Wisconsin breat cancer dataset**	997796	59645	514574	4048784
0.5		997796	76545	495675	3844672
0.75		997796	93467	486955	3546584
0.25	**Heart Disease Dataset**	1060	598	16558	2978878
0.5		1060	841	19657	3547842
0.75		1060	888	22378	3884447

From a pair of healthcare datasets, the person's records are extremely protected by the techniques. The end results of maintained item sets for the made a suggested MOWC, and OHUIS protocol is defined in Table 4.3. For the Wisconsin breast cancer cells dataset, 9845696 substances are considered, coming from that 59846 substances

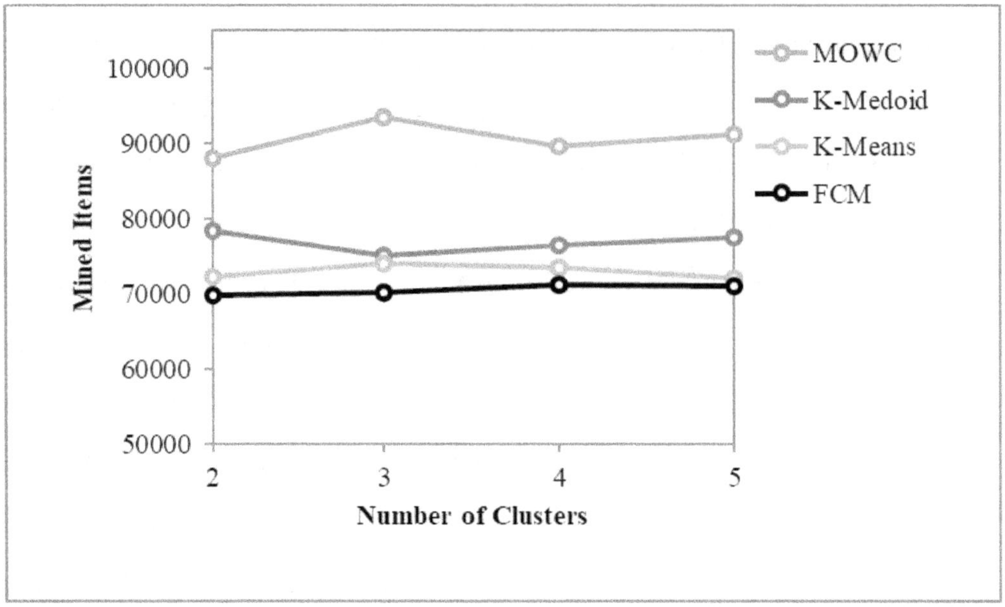

Fig. 4.8 (a) At the threshold level 0.25

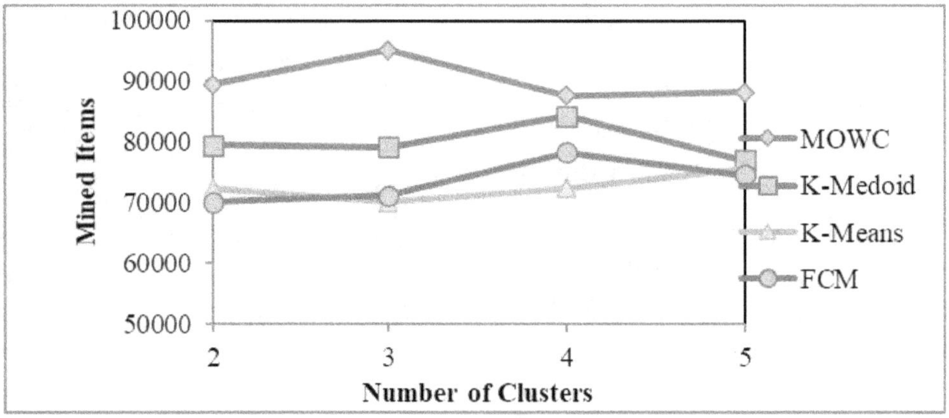

Fig. 4.9 (b) At the threshold level 0.5

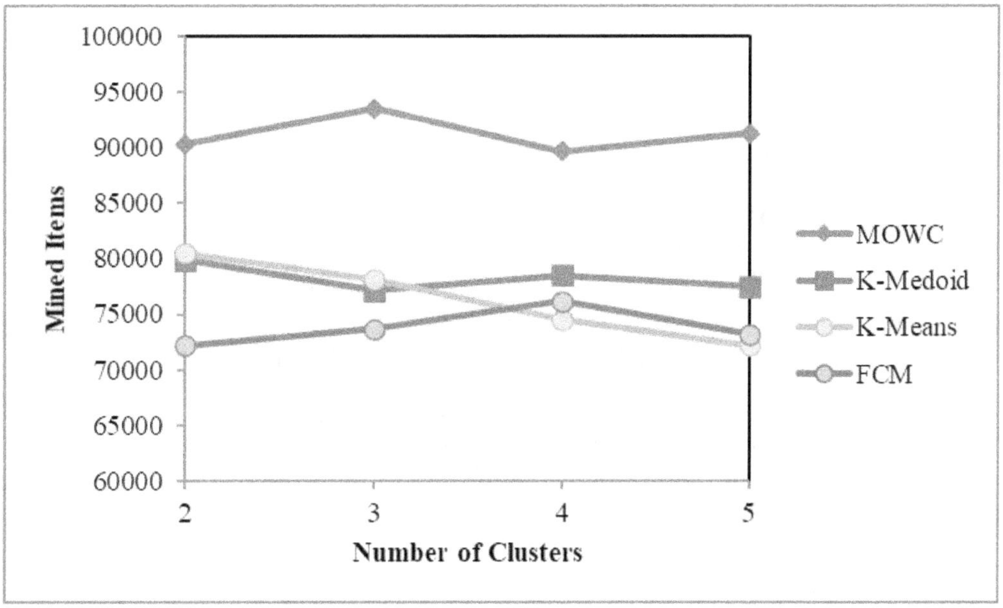

Fig. 4.10 (c) At the threshold level 0.75

are actually preserved, making use of the MOWC concentration and OHUIS exploration protocol. The amount of time taken to preserve information of healthcare is actually 524779 milliseconds, and also the memory allotment is 4047885 bytes. The end results wrap up that if the threshold worth rises, after that, the lot of preserved item sets also improves. The memory allowance, as well as opportunity, is low for the much higher threshold market value. Even more, the relative review of item set exploration formulas is actually explained in body 5, and also, the protocols are actually MOWC, K-mediod, K-means, and FCM.

The above Fig. 4.8 and Fig. 4.9 reveals the clustering precision evaluation of the Wisconsin dataset of breast cancer and also dataset of heart disease. At this juncture, the healthcare information is gathered in various groups using the multi-objective weighted clustering formula. The efficiency of recommended MOWC concentration is examined and compared to the existing FCM and k-medoid protocols. The highest possible clustering precision is actually attained in the MOWC technique is 94 percent aimed at the Wisconsin dataset to breast cancer and also 95 percent for dataset of heart disease at set number 5.

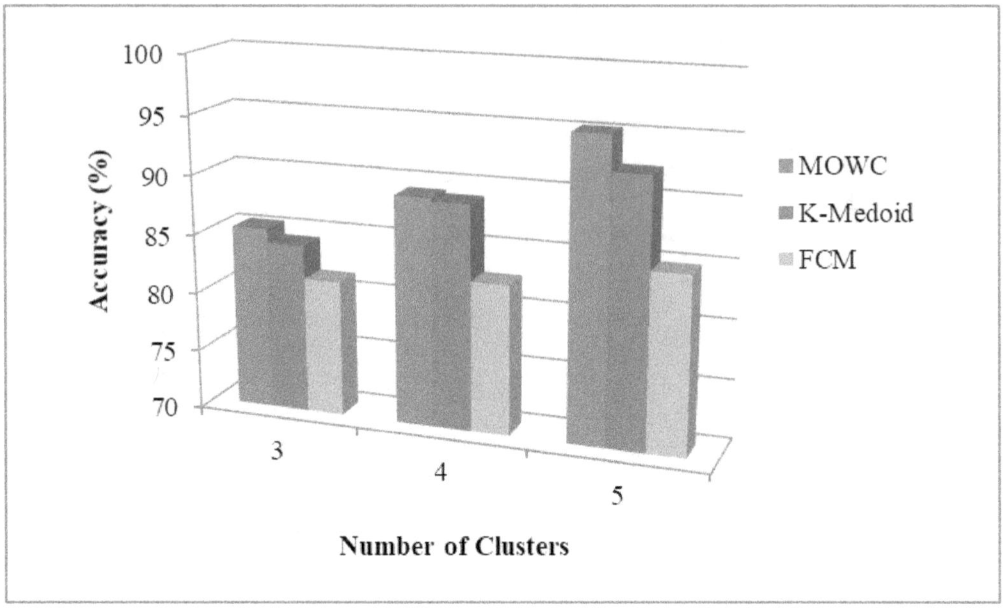

Fig. 4.11 The Wisconsin Breast Cancer Dataset

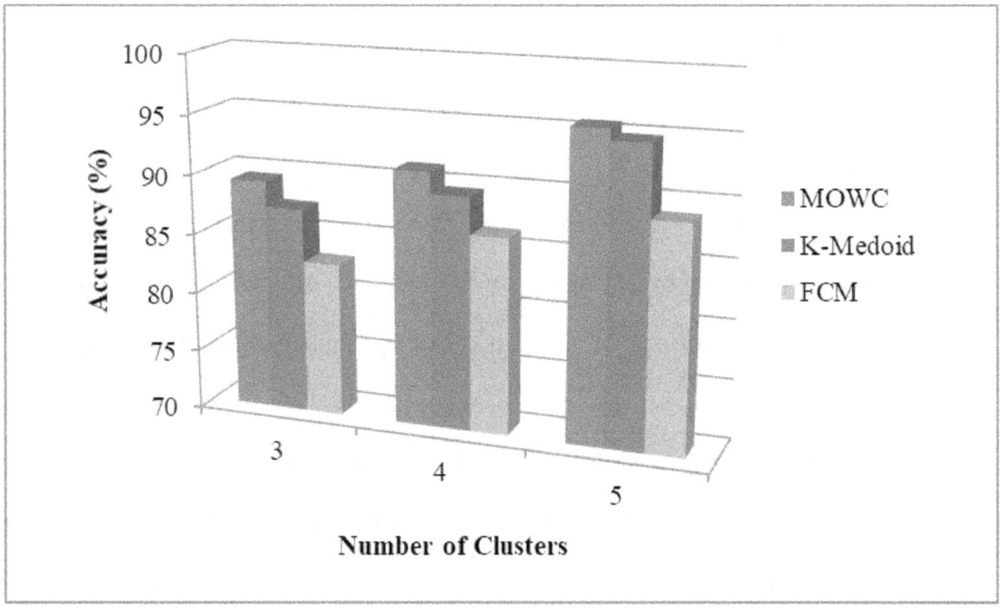

Fig. 4.12 The Heart Disease Dataset

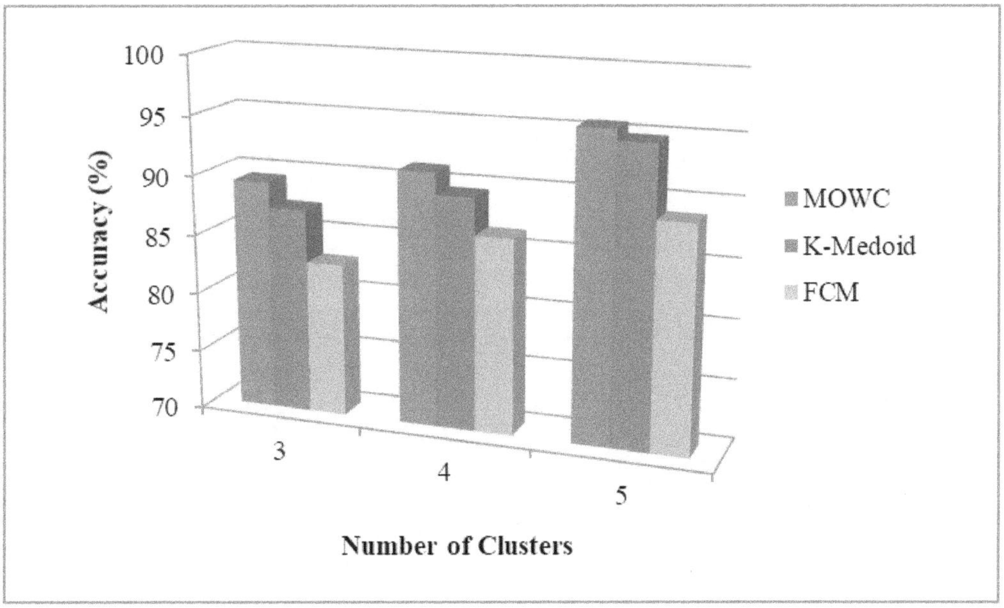

Fig. 4.13 The Wisconsin Breast Cancer Dataset

4.7.1 Running Time and Accuracy Analysis

The moment taken for the item set exploration protocol of Wisconsin data set of breast cancer as well as dataset of heart disease is actually shown in Fig. 4.10 and Fig. 4.11. The operating time is studied for pair of protocols that are actually: KMC-PMUIM and also MOWC-OHUIS item set exploration formulas. While comparing those formulas, the designed MOWC-OHUIS formula accomplishes much less time over other techniques.

Fig. 4.12 reveals the precision analysis of 3 various protocols like MOWC+SS_PSO, K-medoid along with MWO as well as K-medoid with PSO. When thinking about dataset of Wisconsin breast cancer, MOWC with SS_PSO obtains 95.28 percent accuracy, MOWC obtains 94.23 percent reliability, as well as K-medoid along with MWO obtains 92.15 percent, as well as K-medoid with PSO, performs 89.23 percent. For cardiovascular disease dataset, MOWC along with SS_PSO accomplishes 96.18 percent reliability, MOWC obtains 94.78 percent accuracy, and also K-medoid along with MWO obtains 94.19 percent, and also K-medoid along with PSO achieves 85.42 percent. The inspection determines the planned MOWC along with SS_PSO style attained the greatest reliability in preserving the medical care item sets.

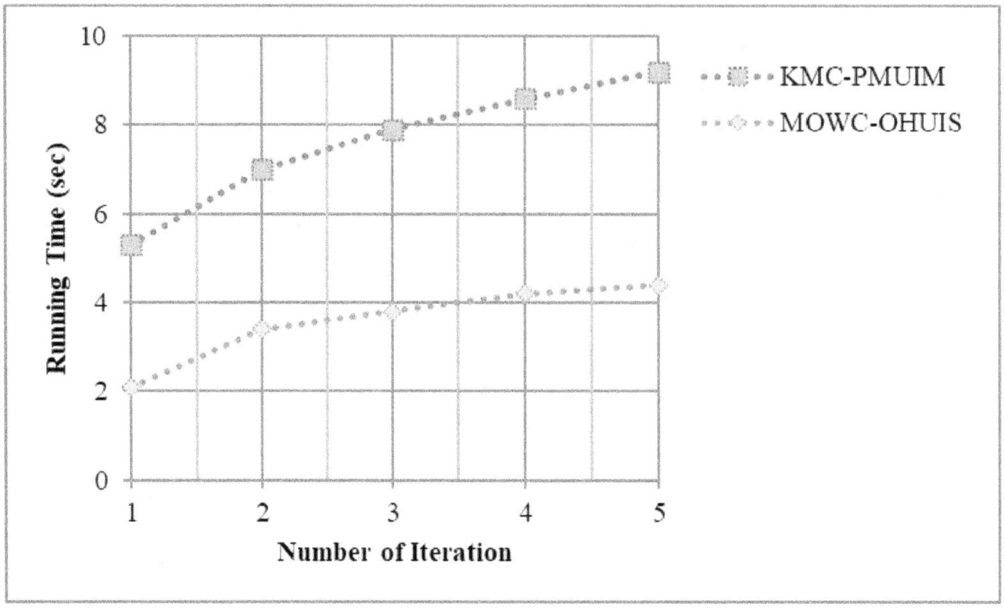

Fig. 4.14 The Heart Disease Dataset

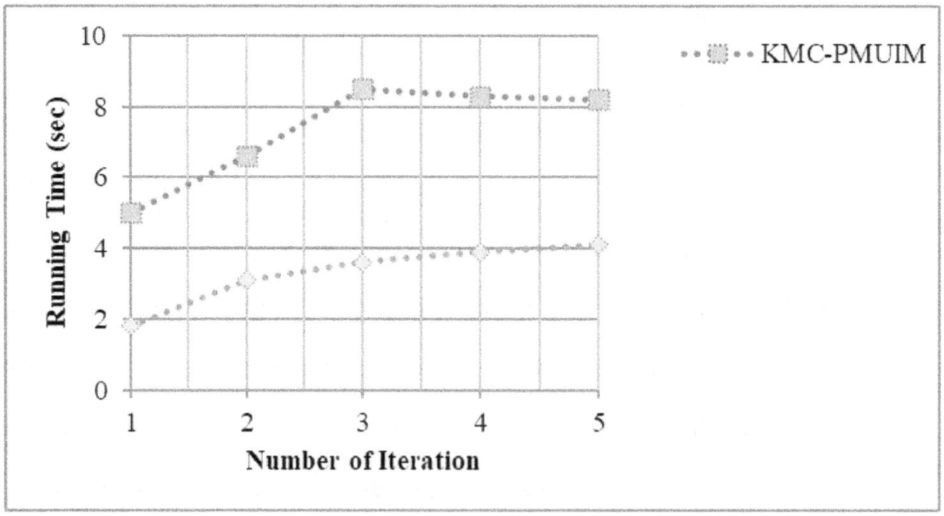

Fig. 4.15 Accuracy Analysis

CHAPTER 5

CONCLUSION

5.1 Conclusion

The enhancing appeal of healthcare communities elevates numerous privacy and also safety problems. These problems are extra serious in preferred social neighborhoods, in which individuals exclusive and delicate info can be just grabbed without user expertise. These communities use open platform communication for all kinds of individuals. Nonetheless, comparable accounts need to be grouped to recognize the feelings of other similar individuals and also to enable communication between them without revealing their exclusive info. The second chapter forecasts an adaptive e-healthcare system by an utilizing k-centroid multi-view point clustering plan. The individual's accounts are separated into client and also caretaker accounts to expand support, also when individuals are unable to gain access to this system. The k-centroid multi-view point clustering method is used to calculate the similarity index and also validate the e-profiles as necessary. Ultimately, all e-profiles are clustered making use of a step-by-step clustering method. The distributed hashing method is utilized to secure the accounts to protect people' delicate info. The speculative results demonstrate that the KCMVS achieves greater accuracy and also lower latency than other well-known privacy preserve clustering strategies.

In third chapter, data mining and medical data security were addressed in the context of e-Health care. Confidentiality and privacy of health information is the biggest obstacle to medical data security. In this research, two medical datasets were taken into consideration for the analysis of e-Health care data while maintaining privacy. K-medoid clustering grouped the data with the smallest distance measurements. Using the PMUIM technique, the patient's itemsets are retained, even when they are clustered. To find the ideal threshold value in PMUIM, the MWO algorithm will have to rely on the hunting behavior of humpback whales. The MWO algorithm is tasked with finding the ideal threshold value, and this will be done based on the behavior of humpback whales. This MWO method chooses the best threshold value for PMUIM, and also increased the privacy accuracy of the itemsets in the two datasets. Additionally, performance over the current privacy-preserving techniques is shown to be improved. By employing itemset mining and hybrid metaheuristic optimization, we increase the medical dataset's privacy in the cloud.

The fourth chapter analyzed the issue of privacy-preserving medical care data re-appropriating. A body based on the crossbreed cloud has popped the question to offer privacy security among large-dimensional medical care details. Under the system, our proposed model contrived 2 healthcare datasets to make them anonymous on the personal as well as public clouds relying on item set mining of random nature. The client's information with associated concerns was actually flocked through making use of the MOWC formula. Then, the organized patient's item sets are secured by making use of the algorithm of OHUIS. In OHUIS, the superior limit market value is actually opted for by the designed SS_PSO algorithm, which implements depending on the abounding actions of salps; it will optimize the value of threshold of the item set mining job. The end results showed that the designed protocol can improve the limit worth of OHUIS as well as strengthen the item sets personal privacy precision in each informational compilations and also its own algorithm running opportunity presents recommended implementation over the other privacy-protecting algorithms.

Finally, there should certainly be a point through factor of privacy suggestions on the solutions as well as the process of e-Health with the target that individuals can feel a lot secured while divulging their health-related details. For future assessments, our experts are going to think of the medical care dataset's privacy in cloud-dependent records resemblance actions by using the item set mining method along with a crossbreed meta-heuristic marketing method.

5.2 Future Scope

The future directions with regard to second chapter consists of proposing a novel similarity measure in terms of weighted calculation of various similarity measures or thinking in the direction of reducing the computation time for calculating the value of Multi-View Similarity measure. Also,we can apply feature extraction techniques to reduce the number of dimensions and to finally reduce the time taken for computing the multi-view similarity measure.

The future directions with regard to third chapter consist of proposing novel clustering techniques in terms of hybrid nature to increase accuracy of clustering. With regard to itemset mining, we may propose novel algorithms to optimize accuracy of privacy preservation in terms of developing algorithms of hybrid nature.

The future directions with regard to fourth chapter consist of proposing novel versions of multi-objective weighted clustering which may gain low computation time and high accuracy. With regard to sensitive itemset mining,we may propose hybrid optimization algorithms in terms of achieving high accuracy for privacy and gaining less time complexity.

REFERENCES

Aamot, H., Kohl, C. D., Richter, D. and Knaup-Gregori, P. (2013), 'Pseudonymization of patient identifiers for translational research', *BMC medical informatics and decision making* **13**(1), 1–15.

Abul-Husn, N. S. and Kenny, E. E. (2019), 'Personalized medicine and the power of electronic health records', *Cell* **177**(1), 58–69.

Adanov, K., Suman, R. S., Gulnara, M. and Akhmaral, K. (2019), 'A short view on the backdrop of american's literature', *Journal of Advanced Research in Dynamical and Control Systems* **11**(12), 182–192.

Agrawal, R., Kiernan, J., Srikant, R. and Xu, Y. (2003), Implementing p3p using database technology, *in* 'Proceedings of the 19th International Conference on Data Engineering (Cat. No. 03CH37405)', IEEE, pp. 595–606.

Alam, S., Dobbie, G., Koh, Y. S., Riddle, P. and Rehman, S. U. (2014), 'Research on particle swarm optimization based clustering: a systematic review of literature and techniques', *Swarm and Evolutionary Computation* **17**(8), 1–13.

Alhaqbani, B. and Fidge, C. (2008), Privacy-preserving electronic health record linkage using pseudonym identifiers, *in* 'Proceedings of the 10th International Conference on e-health Networking, Applications and Services', IEEE, pp. 108–117.

Altamimi, A. M. (2016), 'Security and privacy issues in ehealthcare systems: towards trusted services', *International Journal of Advanced Computer Science and Applications* **7**(9), 229–236.

Andersen, A. and Saus, M. (2017), 'Privacy preserving distributed computation of community health research data', *Procedia computer science* **113**, 633–640.

Banaee, H., Ahmed, M. U. and Loutfi, A. (2013), 'Data mining for wearable sensors in health monitoring systems: a review of recent trends and challenges', *Sensors* **13**(12), 17472–17500.

Bayardo, R. J. and Agrawal, R. (2005), Data privacy through optimal k-anonymization, *in* 'Proceedings of the International conference on data engineering (ICDE'05)', IEEE, pp. 217–228.

Boddy, A., Hurst, W., Mackay, M., El Rhalibi, A., Baker, T. and Montañez, C. A. C. (2019), 'An investigation into healthcare-data patterns', *Future Internet* **11**(2), 1–23.

Cebul, R. D., Love, T. E., Jain, A. K. and Hebert, C. J. (2011), 'Electronic health records and quality of diabetes care', *New England Journal of Medicine* **365**(9), 825–833.

Chen, Y.-R., Rezapour, A. and Tzeng, W.-G. (2018), 'Privacy-preserving ridge regression on distributed data', *Information Sciences* **451**(7), 34–49.

Christy, A., Gandhi, G. M. and Vaithyasubramanian, S. (2015), 'Cluster based outlier detection algorithm for healthcare data', *Procedia Computer Science* **50**, 209–215.

Cox, L. H. (1980), 'Suppression methodology and statistical disclosure control', *Journal of the American Statistical Association* **75**(370), 377–385.

Dalenius, T. (1986), 'Finding a needle in a haystack or identifying anonymous census records', *Journal of official statistics* **2**(3), 329–336.

Dankar, F. K. and Al Ali, R. (2015), 'A theoretical multi-level privacy protection framework for biomedical data warehouses', *Procedia Computer Science* **63**, 569–574.

de Assunção, M. D., di Costanzo, A. and Buyya, R. (2010), 'A cost-benefit analysis of using cloud computing to extend the capacity of clusters', *Cluster Computing* **13**(3), 335–347.

Deng, Y., Jiang, X. and Long, Q. (2020), Privacy-preserving methods for vertically partitioned incomplete data, *in* 'Proceedings of the AMIA Annual Symposium', Vol. 2020, American Medical Informatics Association, pp. 348–357.

El-Gazzar, R., Hustad, E. and Olsen, D. H. (2016), 'Understanding cloud computing adoption issues: A delphi study approach', *Journal of Systems and Software* **118**(8), 64–84.

Fernández-Alemán, J. L., Señor, I. C., Lozoya, P. Á. O. and Toval, A. (2013), 'Security and privacy in electronic health records: A systematic literature review', *Journal of biomedical informatics* **46**(3), 541–562.

Forman, G. and Zhang, B. (2000), 'Distributed data clustering can be efficient and exact', *ACM SIGKDD explorations newsletter* **2**(2), 34–38.

Fung, B. C., Wang, K., Fu, A. W.-C. and Pei, J. (2008), Anonymity for continuous data publishing, *in* 'Proceedings of the 11th international conference on Extending database technology: Advances in database technology', pp. 264–275.

Giniat, E. J. (2011), 'Cloud computing: innovating the business of health care: healthcare finance executives can lead their organizations in exploring ways to use cloud computing for operational success', *healthcare financial management* **65**(5), 130–132.

Gøeg, K. R., Cornet, R. and Andersen, S. K. (2015), 'Clustering clinical models from local electronic health records based on semantic similarity', *Journal of biomedical informatics* **54**, 294–304.

Goksen, Y., Eminagaoglu, M., Dogan, O. *et al.* (2011), 'Data mining in medical records for the enhancement of strategic decisions: a case study', *Scientific Bulletin–Economic Sciences* **10**(1), 135–45.

Guha, S., Rastogi, R. and Shim, K. (2001), 'Cure: an efficient clustering algorithm for large databases', *Information systems* **26**(1), 35–58.

Haldorai, A. and Kandaswamy, U. (2019), Supervised machine learning techniques in intelligent network handovers, *in* 'Intelligent Spectrum Handovers in Cognitive Radio Networks', Springer, pp. 135–154.

Hamza, N. and Hefny, H. A. (2013), 'Attacks on anonymization-based privacy-preserving: a survey for data mining and data publishing', **4**(2), 101–112.

Harikumar, S. and Surya, P. (2015), 'K-medoid clustering for heterogeneous datasets', *Procedia Computer Science* **70**, 226–237.

Henze, M., Hermerschmidt, L., Kerpen, D., Häußling, R., Rumpe, B. and Wehrle, K. (2016), 'A comprehensive approach to privacy in the cloud-based internet of things', *Future Generation Computer Systems* **56**, 701–718.

Huang, L.-C., Chu, H.-C., Lien, C.-Y., Hsiao, C.-H. and Kao, T. (2009), 'Privacy preservation and information security protection for patients' portable electronic health records', *Computers in Biology and Medicine* **39**(9), 743–750.

Islam, M., Rahman, A., Akhter, H., Begum, M., Mawla, Y., Asaduzzaman, K. and Hoq, M. (2015), 'Microcontroller based constant voltage maximum power point tracking for solar inverter applications', *International Journal of Advanced Engineering, Management and Science* **1**(9), 11–14.

Karaboga, D. (2005), An idea based on honey bee swarm for numerical optimization, Technical report, Erciyes university, engineering faculty, computer engineering department.

Khan, A. R. (2012), 'Access control in cloud computing environment', *ARPN Journal of Engineering and Applied Sciences* **7**(5), 613–615.

Khanna, S., Kumari, S. and Tadir, T. (2015), 'Comprehensive review of huffman encoding technique for image compression', *International Journal of Advanced Engineering, Management and Science* **1**(9), 8–10.

Kumar, R. (2014), 'Design and implementation of intelligent digital home thermostat using nxp low power 32bits lpc2148 microcontroller', *International Journal of Advanced Engineering Research And Science* **1**(2), 57–60.

LeFevre, K., DeWitt, D. J. and Ramakrishnan, R. (2005), Incognito: Efficient full-domain k-anonymity, *in* 'Proceedings of the ACM SIGMOD International Conference on Management of data', pp. 49–60.

Li, D., Lv, Q., Shang, L. and Gu, N. (2017), 'Efficient privacy-preserving content recommendation for online social communities', *Neurocomputing* **219**(1), 440–454.

Li, N., Li, T. and Venkatasubramanian, S. (2007), t-closeness: Privacy beyond k-anonymity and l-diversity, *in* 'Proceedings of the IEEE 23rd International Conference on Data Engineering', IEEE, pp. 106–115.

Lin, C.-W., Abdul, S. S., Clinciu, D. L., Scholl, J., Jin, X., Lu, H., Chen, S. S., Iqbal, U., Heineck, M. J. and Li, Y.-C. (2014), 'Empowering village doctors and enhancing rural healthcare using cloud computing in a rural area of mainland china', *Computer methods and programs in biomedicine* **113**(2), 585–592.

Loukides, G., Liagouris, J., Gkoulalas-Divanis, A. and Terrovitis, M. (2014), 'Dis-association for electronic health record privacy', *Journal of biomedical informatics* **50**(3), 46–61.

Machanavajjhala, A., Kifer, D., Gehrke, J. and Venkitasubramaniam, M. (2007), 'l-diversity: Privacy beyond k-anonymity', *ACM Transactions on Knowledge Discovery from Data (TKDD)* **1**(1), 1–24.

Majeed, A. (2019), 'Attribute-centric anonymization scheme for improving user privacy and utility of publishing e-health data', *Journal of King Saud University-Computer and Information Sciences* **31**(4), 426–435.

Meingast, M., Roosta, T. and Sastry, S. (2006), Security and privacy issues with health care information technology, *in* 'Proceedings of the International Conference of the IEEE Engineering in Medicine and Biology Society', IEEE, pp. 5453–5458.

Meri, A., Hasan, M., Danaee, M., Jaber, M., Safei, N., Dauwed, M., Abd, S. K., Albsheish, M. *et al.* (2019), 'Modelling the utilization of cloud health information systems in the iraqi public healthcare sector', *Telematics and Informatics* **36**(1), 132–146.

Metsker, O., Bolgova, E., Yakovlev, A., Funkner, A. and Kovalchuk, S. (2017), 'Pattern-based mining in electronic health records for complex clinical process analysis', *Procedia computer science* **119**, 197–206.

Michalas, A. and Komninos, N. (2014), The lord of the sense: A privacy preserving reputation system for participatory sensing applications, *in* 'Proceedings of the IEEE Symposium on Computers and Communications (ISCC)', IEEE, pp. 1–6.

Mills, S. (2019), 'Electronic health records and use of clinical decision support', *Crit Care Nurs Clin North Am* **31**(2), 125–131.

Miotto, R., Li, L., Kidd, B. A. and Dudley, J. T. (2016), 'Deep patient: an unsupervised representation to predict the future of patients from the electronic health records', *Scientific reports* **6**(1), 1–10.

Mirjalili, S., Gandomi, A. H., Mirjalili, S. Z., Saremi, S., Faris, H. and Mirjalili, S. M. (2017), 'Salp swarm algorithm: A bio-inspired optimizer for engineering design problems', *Advances in Engineering Software* **114**, 163–191.

Monegai, B. (2013), 'Ge healthcare launches cloud service', *Healthcare IT News* .
URL: *http://www.healthcareitnews.com/news/ge-healthcare-launches-cloudservice*

Muller, E., Gunnemann, S., Farber, I. and Seidl, T. (2012), Discovering multiple clustering solutions: Grouping objects in different views of the data, *in* 'Proceedings of the IEEE 28th International Conference on Data Engineering', IEEE, pp. 1207–1210.

Nadaf, S. J. and Patil, R. (2016), Cloud based privacy preserving secure health data storage and retrieval system, *in* 'Proceedings of the International Conference on Inventive Computation Technologies (ICICT)', Vol. 2, IEEE, pp. 1–6.

Nayahi, J. J. V. and Kavitha, V. (2017), 'Privacy and utility preserving data clustering for data anonymization and distribution on hadoop', *Future Generation Computer Systems* **74**, 393–408.

Nguyen, D. T., Chen, L. and Chan, C. K. (2011), 'Clustering with multiviewpoint-based similarity measure', *IEEE transactions on knowledge and data engineering* **24**(6), 988–1001.

Ogbuabor, G. and Ugwoke, F. (2018), 'Clustering algorithm for a healthcare dataset using silhouette score value', *International Journal of Computer Science and Information Technology (IJCSIT)* **10**(2), 27–37.

Otair, M. (2013), 'Approximate k-nearest neighbour based spatial clustering using k-d tree', *International Journal of Database Management Systems* **5**, 97–108.

Palarivattom, S. and Kochunni, K. (2015), 'Real-time, interposable communication for web services', *International Research Journal of Management, IT and Social Sciences* **2**(5), 26–32.

Pan, W.-T. (2012), 'A new fruit fly optimization algorithm: taking the financial distress model as an example', *Knowledge-Based Systems* **26**, 69–74.

Pasupuleti, S. K., Ramalingam, S. and Buyya, R. (2016), 'An efficient and secure privacy-preserving approach for outsourced data of resource constrained mobile devices in cloud computing', *Journal of Network and Computer Applications* **64**, 12–22.

Pirbhulal, S., Samuel, O. W., Wu, W., Sangaiah, A. K. and Li, G. (2019), 'A joint resource-aware and medical data security framework for wearable healthcare systems', *Future Generation Computer Systems* **95**, 382–391.

Pussewalage, H. S. G. and Oleshchuk, V. A. (2016), 'Privacy preserving mechanisms for enforcing security and privacy requirements in e-health solutions', *International Journal of Information Management* **36**(6), 1161–1173.

Qamar, N., Yang, Y., Nadas, A. and Liu, Z. (2016), 'Querying medical datasets while preserving privacy', *Procedia Computer Science* **98**, 324–331.

Rahman, F., Bhuiyan, M. Z. A. and Ahamed, S. I. (2017), 'A privacy preserving framework for rfid based healthcare systems', *Future generation computer systems* **72**, 339–352.

Rajest, S. S. and Suresh, D. (2018), 'The deducible teachings of historiographic metafiction of modern theories of both fiction and history', *Eurasian Journal of Analytical Chemistry* **13**(4), 110–117.

Rao, O. S. and Akiri, C. K. (2014), 'Plaintext encryption and decryption using koblitz curves', *International Journal of Advanced Engineering Research and Science (IJAERS)* **1**(2), 1–6.

Reddy, D. and d'Almaine, F. (2015), 'The effectiveness of using a non-platinum catalyst for a proton exchange membrane fuel cell (pemfc)', *International Journal of Advanced Engineering, Management and Science* **1**(9), 1–8.

Riedl, B., Grascher, V., Fenz, S. and Neubauer, T. (2008), Pseudonymization for improving the privacy in e-health applications, *in* 'Proceedings of the 41st Annual Hawaii International Conference on System Sciences (HICSS 2008)', IEEE, pp. 255–255.

Rocha, F., Abreu, S. and Correia, M. (2011), 'The final frontier: Confidentiality and privacy in the cloud', *Computer* **44**(9), 44–50.

Sahi, A., Lai, D. and Li, Y. (2016), 'Security and privacy preserving approaches in the ehealth clouds with disaster recovery plan', *Computers in biology and medicine* **78**, 1–8.

Sahi, M. A., Abbas, H., Saleem, K., Yang, X., Derhab, A., Orgun, M. A., Iqbal, W., Rashid, I. and Yaseen, A. (2017), 'Privacy preservation in e-healthcare environments: State of the art and future directions', *Ieee Access* **6**, 464–478.

Sahney, R. and Sharma, M. (2018), 'Electronic health records: A general overview', *Current Medicine Research and Practice* **8**(2), 67–70.

Samarati, P. (2001), 'Protecting respondents identities in microdata release', *IEEE transactions on Knowledge and Data Engineering* **13**(6), 1010–1027.

Samarati, P. and Sweeney, L. (1998), Generalizing data to provide anonymity when disclosing information, *in* 'Proceedings of Symposium on Principles of Database Systems', Vol. 98, pp. 10–1145.

Sandanasamy, D., Govindarajane, S. and Sundararajan, T. (2014), 'Energy efficient lighting systems in green buildings: An overview', *International Journal of Advanced Engineering Research And Science* **1**(2), 44–50.

Shakeabubakor, A., Sundararajan, E. and Hamdan, A. (2015), 'The integration of cloud computing with lean theory to improve academic researcher's productivity', *International Journal of Advances in Computer Science Its Applications– IJCSIA* **5**, 18–25.

Sharawi, M., Zawbaa, H. M. and Emary, E. (2017), Feature selection approach based on whale optimization algorithm, *in* 'Proceedings of the Ninth international conference on advanced computational intelligence (ICACI)', IEEE, pp. 163–168.

Sharifi, Z., Rezghi, M. and Nasiri, M. (2013), New algorithm for recommender systems based on singular value decomposition method, *in* 'Proceedings of the International eConference on Computer and Knowledge Engineering', IEEE, pp. 86–91.

Shen, N., Bernier, T., Sequeira, L., Strauss, J., Silver, M. P., Carter-Langford, A. and Wiljer, D. (2019), 'Understanding the patient privacy perspective on health information exchange: a systematic review', *International journal of medical informatics* **125**, 1–12.

Singh, A. and Liu, L. (2003), Trustme: anonymous management of trust relationships in decentralized p2p systems, *in* 'Proceedings of the Third International Conference on Peer-to-Peer Computing (P2P2003)', IEEE, pp. 142–149.

Sonnati, R. (2017), 'Improving healthcare using big data analytics', *International Journal of Scientific and Technology Research* **6**(3), 142–146.

Sreedhar, K., Faruk, M. and Venkateswarlu, B. (2017), 'A genetic tds and bug with pseudo-identifier for privacy preservation over incremental data sets', *Journal of Intelligent and Fuzzy Systems* **32**(4), 2863–2873.

Sultan, N. (2014), 'Making use of cloud computing for healthcare provision: Opportunities and challenges', *International Journal of Information Management* **34**(2), 177–184.

Sun, C., Fu, Y., Zhou, J. and Gao, H. (2014), 'Personalized privacy-preserving frequent itemset mining using randomized response', *The Scientific World Journal* **2014**, 1–10.

Sweeney, L. (2002*a*), 'Achieving k-anonymity privacy protection using generalization and suppression', *International Journal of Uncertainty, Fuzziness and Knowledge-Based Systems* **10**(05), 571–588.

Sweeney, L. (2002*b*), 'k-anonymity: A model for protecting privacy', *International Journal of Uncertainty, Fuzziness and Knowledge-Based Systems* **10**(05), 557–570.

Taneja, H., Singh, A. K. *et al.* (2015), 'Preserving privacy of patients based on re-identification risk', *Procedia Computer Science* **70**, 448–454.

Thilakanathan, D., Chen, S., Nepal, S., Calvo, R. and Alem, L. (2014), 'A platform for secure monitoring and sharing of generic health data in the cloud', *Future Generation Computer Systems* **35**, 102–113.

Upmanyu, M., Namboodiri, A. M., Srinathan, K. and Jawahar, C. (2010), Efficient privacy preserving k-means clustering, *in* 'Pacific-Asia Workshop on Intelligence and Security Informatics', Springer, pp. 154–166.

Van Gorp, P., Comuzzi, M., Jahnen, A., Kaymak, U. and Middleton, B. (2014), 'An open platform for personal health record apps with platform-level privacy protection', *Computers in biology and medicine* **51**, 14–23.

Vimalachandran, P., Wang, H., Zhang, Y., Heyward, B. and Whittaker, F. (2016), Ensuring data integrity in electronic health records: a quality health care implication, *in*

'Proceedings of the International Conference on Orange Technologies (ICOT)', IEEE, pp. 20–27.

Vora, J., Italiya, P., Tanwar, S., Tyagi, S., Kumar, N., Obaidat, M. S. and Hsiao, K.-F. (2018), Ensuring privacy and security in e-health records, *in* 'Proceedings of the International conference on computer, information and telecommunication systems (CITS)', IEEE, pp. 1–5.

Wang, W., Chen, L. and Zhang, Q. (2015), 'Outsourcing high-dimensional healthcare data to cloud with personalized privacy preservation', *Computer Networks* **88**, 136–148.

Wang, X., Bai, L., Yang, Q., Wang, L. and Jiang, F. (2019), 'A dual privacy-preservation scheme for cloud-based ehealth systems', *Journal of Information Security and Applications* **47**, 132–138.

Wang, X., Cheng, W., Mohapatra, P. and Abdelzaher, T. (2013*a*), 'Enabling reputation and trust in privacy-preserving mobile sensing', *IEEE Transactions on Mobile Computing* **13**(12), 2777–2790.

Wang, X. O., Cheng, W., Mohapatra, P. and Abdelzaher, T. (2013*b*), Artsense: Anonymous reputation and trust in participatory sensing, *in* 'Proceedings of the IEEE INFOCOM', IEEE, pp. 2517–2525.

Wu, R. (2012), *Secure sharing of electronic medical records in cloud computing*, Arizona State University.

Xiao, X. and Tao, Y. (2007), M-invariance: towards privacy preserving re-publication of dynamic datasets, *in* 'Proceedings of the ACM SIGMOD International Conference on Management of data', pp. 689–700.

Xu, L., Jiang, C., Wang, J., Yuan, J. and Ren, Y. (2014), 'Information security in big data: privacy and data mining', *Ieee Access* **2**, 1149–1176.

Yang, J.-J., Li, J.-Q. and Niu, Y. (2015), 'A hybrid solution for privacy preserving medical data sharing in the cloud environment', *Future Generation computer systems* **43**, 74–86.

Yüksel, B., Küpçü, A. and Özkasap, Ö. (2017), 'Research issues for privacy and security of electronic health services', *Future Generation Computer Systems* **68**, 1–13.

Zhu, X., Tu, S.-P., Sewell, D., Yao, N. A., Mishra, V., Dow, A. and Banas, C. (2019), 'Measuring electronic communication networks in virtual care teams using electronic health records access-log data', *International journal of medical informatics* **128**, 46–52.

www.ingramcontent.com/pod-product-compliance
Lightning Source LLC
LaVergne TN
LVHW082343200926
848481LV00012B/3437
* 9 7 9 8 2 3 2 0 5 4 7 1 7 *